SPACE FACTS

FOR CURIOUS MINDS

*5,705 Astronomy Facts About Astronauts, Planets, Galaxies &
Everything in Between with Fun Trivia*

PANTHEON SPACE ACADEMY

GET MORE WITH THESE QR CODES

My Amazon Author Page - https://amazon.com/author/xyz

Free resources for you on our website -
https://www.pantheonspace.com

Join us on Facebook to keep the conversation going -
https://www.facebook.com/pantheonspace

TABLE OF CONTENTS

FUN FACTS
SPACE TRIVIA

PANTHEON SPACE ACADEMY

INTRODUCTION

The dawn of Space travel is here. Rocket technology will soon welcome entire families onboard for the ultimate adventure. As the Space race heats up, so does the information age. We are the future of Space reporting. Say goodbye to heavy encyclopedias and boring lectures trying to conduct the wonders of the Universe. It's time for Outer Space to get delivered to you with style and charisma! At Pantheon Space Academy, we know a lot about astronomy, and it's always fun to put that experience to the test!

Welcome to *Fun Facts Space Trivia*! Whether you want to learn more about Space for yourself or bring excitement and creativity to your time with friends and family, this book is for you. Together we can experience the fascination of Space discovery one trivia question at a time! Part one book has 177 trivia questions to challenge everybody, from students to novice astronomers and above. You then receive the answer with multiple facts on the subject. In this book, you'll find Space facts about planets, moons, stars, black holes, and more!

Grab a notepad to keep score or just read along to learn more. You'll have a fun time expanding your knowledge of the extraordinary and ever-expanding Universe. Without further ado, let's gather the crew, suit up, and blast off into the beautiful cosmos!

CHAPTER ONE

OUR PLANETS AND MOONS

According to NASA, an object has to (1) orbit a star, (2) be spherical, and (3) be the dominant body in its orbital neighborhood to be recognized as what?

A planet. This definition was officially adopted in 2006 by the International Astronomical Union due to newly detected objects which fit the first two requirements but didn't feel like authentic planets.

How many planets are there in our Solar System?

Eight. Sorry, Pluto: in 2006, when the new definition of "planet" was adopted, Pluto was reclassified as a "dwarf planet" because it was not the dominant body in its orbital neighborhood. There were objects of comparable size nearby.

What is the natural satellite of a planet called?

Most commonly referred to as a moon. A "natural satellite" means that it was not man-made and orbits another object such as a planet, minor planet, or other Solar System body. Any natural satellite of another object is a moon unless that other object is a star. In that case, the orbiting object is either a planet, a dwarf planet, or an asteroid.

Is it possible for something other than a planet to have a moon?

Yes! The dwarf planet Pluto has five of them. Its most prominent moon, Charon, almost acts to create a binary system between it and Pluto, though Pluto is the slightly dominant object. Even large asteroids can have moons of their own!

Although most often called just "the Moon," what is the second-most-common name for Earth's only natural satellite?

Luna. That is the Latin term for Earth's only natural satellite, and it's also where the word "lunar" comes from; it always refers to anything related to our Moon.

How do astronomers think the Moon originated?

The most popular idea is that a Mars-sized object collided with Earth about 4 billion years ago, and the material ejected from this impact became the Moon. The hypothetical object's name is "Theia," the mother of Greek mythology's moon goddess.

In which country was the largest known meteor impact?

South Africa. The meteor in question caused the Vredefort crater, which is 186 miles across. For perspective, that is twice the size of the Chicxulub crater in the Yucatán Peninsula, created by the meteor that likely killed the dinosaurs!

Which two planets do not have any moons?

Mercury and Venus. As these are the two planets closest to the Sun, any celestial body too far away would be pulled into the Sun, but the planet's gravity would destroy anything close enough. There is no middle ground with these.

Name the largest moon in our Solar System?

Ganymede. The moon of Jupiter is so big that it's about 8% larger than our tiniest planet, Mercury. Of course, the fact that Ganymede orbits a planet disqualifies it from being a planet itself.

Which gas giant has a moon bulkier than Mercury?

Saturn and the moon in question is Titan. Not only is Titan larger in size than Mercury, however, it also has something the tiniest planet doesn't: a substantial atmosphere. That makes it unique among all known moons in the Solar System.

Which planet is the only one named after a Greek god?

Uranus. The god Uranus's son, Cronus, is the Greek equivalent of the Roman god Saturn, and Cronus's sons Zeus and Poseidon are the Greek equivalents of Jupiter and Neptune, the other three giant planets. Uranus's Roman equivalent is Caelus.

Name the color that primarily makes up sunrises and sunsets on Mars.

Blue! As the Sun is closer to the horizon, dust particles on Mars filter blue from sunlight more efficiently than other colors. Later in the day, though, the sky gains more yellows and reds.

How many Earth years (about 365 days) does it take for Neptune to orbit the Sun?

Almost 165, nearly double its neighbor Uranus's 84-year orbit! Astronomers discovered Neptune in 1846, and in 2011, the windy globe completed its first revolution since being identified. Neptune will complete the second revolution in 2176.

Which planet has the nickname "Red Planet"?

Mars. Named for its distinctive, dusty-red color. This appearance is due to the abundance of iron oxide on the Martian surface. That is the same compound that creates rust on metal.

If you had a pool wide enough, one of our planets could float on top! Name the planet with the least density?

Saturn is less dense than any of our planetary neighbors. The planet with the highest density is Earth.

How many planets have rings?

Four: Jupiter, Saturn, Uranus, and Neptune. Saturn's rings, however, are by far the most prominent, and because of this, they're the only ones commonly depicted in images of their planet. (Some depictions of Uranus do show its rings, but that's far less common.)

What planet lays on its side as it orbits the Sun?

Uranus! Its poles lay where the equator would be on most other planets. This tilt is likely from a collision with an Earth-sized object earlier in its history. The angle also means that its seasons last for 21 Earth years.

At night during a clear sky, how many planets can you see without using a telescope?

Five: Mercury, Venus, Mars, Jupiter, and Saturn. Amateurs and astronomers have watched these for millennia. Uranus was the first planet to be discovered with a telescope and officially found in 1781. Neptune was predicted to exist, thanks to mathematical calculations, before its detection in 1846.

Terrestrial planets are made mostly of rocky and metallic substances. There are four in our Solar System. Can you name them?

Mercury, Venus, Earth, and Mars. Jupiter and Saturn are "gas giants" because they are primarily hydrogen and helium. At the same time, Uranus and Neptune are "ice giants" because their makeup is of water, methane, and ammonia.

In terms of their rotation on their axis, what do Venus and Uranus have in common?

They rotate clockwise. All other planets in the Solar System rotate counterclockwise on their axis, including Earth. Rotation makes a day on Earth, while a revolution around the Sun makes a year. All planets revolve around the Sun in the same direction.

Name the most volcanically active body in the Solar System.

Io. The moon of Jupiter is the only place outside Earth where active volcanoes are known to exist. They may live on Venus, but its atmosphere blocks our view, while Mars may have had active volcanoes in its past.

Aside from Earth, what other planet has a formation called "Mount Olympus"?

Mars. The formation's name is Olympus Mons, Latin for "Mount Olympus." Two and a half times the elevation of our very own Mt. Everest. Mariner 9 first seen what is now known as the tallest planetary mountain in our Solar System. Fitting for the seat of the Greek gods!

Which planet is 2.5 times as massive as all the other spheres combined?

Jupiter. Yes, all other planets combined can't rival the mass of Jupiter. Still, to put it in perspective, it would require more than one thousand Jupiters to fill up the Sun! Even giant Jupiter bows to the Sun's enormous presence and powerful gravity.

Two planets are competing to have the most moons. Which planet is in the lead?

As of 2021, Saturn has 53 known moons, with 29 of those pending confirmation. Jupiter, far bulkier than Saturn, also has 53 known moons, but with only 26 awaiting validation, Saturn is likely to be the one with more moons.

True or false: Earth is almost double the size of Mars?

True! While Venus and Earth are about the same size, and Mercury is by far the littlest planet, Mars is almost half the size of Earth. Because of this, if you weighed 200 pounds on Earth, you'd only weigh 75 on Mars!

Which planet's moons have names from characters of literature, as opposed to mythology?

Uranus. The natural satellite names are characters from William Shakespeare's plays, like Juliet, Ophelia, and Puck, with a few named after figures from another English writer, Alexander Pope.

Phobos, the Greek god of horror, orbits which planet?

Mars. Continuously escorting Phobos is a second Martian moon, Deimos, whose Greek namesake is the twin embodiment of dread. The two gods were actual twins, their two emotions associated with war, and since Mars was the god of war, astronomers chose their names, respectively.

Which planet has wind speeds faster than the speed of sound?

Neptune. Wind speeds on the ice giant can reach 1,600 miles per hour. That's three times faster than a commercial airplane. The fastest hurricanes on Earth travel at just about a *tenth* of that speed!

Which two planets have massive storms on their surface that you can see through a telescope?

Jupiter and Neptune. In every depiction, we can see Jupiter's prominent Great Red Spot. You can even fit Earth inside the colossal storm. Neptune has a similar anticyclonic pattern dubbed the Great Dark Spot, discovered in 1989 by NASA's Voyager 2 Space probe.

True or false: moons always remain stable in their composition and size?

False! While this is true for large moons, including Luna, Saturn's rings are home to many small moons floating amidst material within the rings. Astrophysicists aren't even sure the moons will remain together in the long run.

True or false: the Moon is slowly drifting away from Earth?

True! Over the few billion years that the Moon has been a part of Earth's orbit, it has slowly begun to drift away from us. However, don't worry! The drift is only 1.5 inches per year, so it would take 42,240 years to move an entire mile.

Which is the only planet not named after any god?

Earth. Like "the Moon," the word for Earth comes from an Old English word, this one meaning "ground." That is why "earth" when lowercase refers to dirt. Our globe was thoughtfully named because it didn't seem like just another planet to ancient peoples.

Creatively this fairy-tale-inspired name was given to the ideal zone a planet needs to support life?

The Goldilocks Zone. Her character tried three bowls of porridge, one too hot, one too cold, and one just right. Planets too close to the star would be too hot for supporting life but too far away, and the Earth would be too cold and bleak for life.

What is the most common element in the Martian atmosphere?

Carbon dioxide. The atmosphere is only 1% as thick as Earth's, composed almost entirely of Carbon dioxide. CO_2 is a compound made of two elements from the Periodic Table. This compound is vital in the search for life beyond our planet. In a recent toaster-sized experiment, scientists succeeded in finding a process that would convert CO_2 to oxygen, so astronauts on Mars could produce their air supply there.

Which component of Earth protects it from solar storms and causes the Northern Lights?

The magnetosphere. While not the strongest in the Solar System, Earth has a relatively robust magnetic layer. The activity at the poles redirects our invisible invaders; this energy interaction with the atmosphere causes a memorable sight called the Northern Lights, otherwise known as the Aurora Borealis.

They orbit gas giants; which two moons receive the most attention as possibly supporting life?

Europa and Titan. One from Saturn and the other from Jupiter. Europa is perhaps the most promising candidate due to its likelihood of having an underground ocean. However, Titan is one of the few objects in the entire Solar System with an atmosphere, an integral component known to support surface life.

Which planet likely once had liquid water on its surface?

Mars. The Red Planet still possesses some water in the form of ice, but billions of years in its past, water was probably abundant on its surface, along with it having a thicker atmosphere. Those may have been enough for microbial life to form, but Mars did not remain hospitable to life.

Which moon of Jupiter is the most cratered object known in the Solar System?

Callisto. Close in size to Mercury and thereby the third-largest moon in the Solar System. Callisto may also potentially have an underground ocean, like its sister moon Europa, but likely buried much deeper making Callisto a less-promising place for life to form.

What planet takes two trips around the Sun for every three spins on its axis?

Mercury. Although this pattern likely didn't start with Mercury's actual formation, it did become locked in quite early for the Swift Planet. This type of phenomenon is "resonance," this is a periodic gravitational influence that two celestial bodies have on each other. There are many types of resonance.

Which celestial body is about five times further away from the Sun as Earth is?

Jupiter. Because astronomers have known this for centuries, accurately calculating the distance from Earth to the Sun would allow them to calculate Jupiter's distance from both objects.

Which two planets lend their names to elements on the periodic table?

Uranus and Neptune, for uranium and neptunium, respectively. When plutonium, the element after neptunium, was given its name, Pluto had only recently been discovered and still considered a planet, so it fitted with the pattern.

True or false: the presence of massive impact craters on a planet or moon indicates that it has not been geologically active for a long time.

True! A lack of geological activity allows these impact craters to remain on the surface, and the presence of a lot of them on bodies such as the Moon and Mercury indicates that these bodies have been inactive for most of their history since they finished forming.

Outside of Earth, what is the most likely planet in the Solar System to harbor life?

Venus! There is a narrow band in its atmosphere where temperatures may range between 30 and 200 degrees Fahrenheit. While far from an ideal place for life, astronomers announced, in 2020, that the presence of phosphine gas indicates a chance that life may already be on our sister planet.

Galileo directly observed which planet's four biggest moons?

Jupiter. Discovered in 1610 and named in 1614, the four moons were Io, Europa, Ganymede, and Callisto, which he suggested calling simply Jupiter I–IV. Although the mythological names won out, his system still categorizes moons concerning newly discovered ones for any planet.

Aside from Earth, what other planet in the Solar System could be called the "Blue Planet" due to its blue appearance?

Neptune. Named after the Roman god of the sea. Sequentially to keep in accord with the terms of the other planets in the Solar System, apart from Earth. Neptune's blue appearance occurs from methane in its outer atmosphere.

What would it mean if something were "Jovian"?

It would mean that it had come from, or was related to, the planet Jupiter. For instance, a Jovian year is 11.86 Earth years. The term comes from "Jove," a synonym for the god Jupiter which shares the same linguistic origin as "Jupiter."

Into what planet's atmosphere was the "Cassini" probe intentionally sent, where it then vaporized?

Saturn. "Cassini" explored the gas giant from 2004 to 2017, orbiting it 294 times within those thirteen years. Knowing "Cassini" was at the end of its lifespan, scientists running the mission sent it into the planet to gather and transmit close-up data before its demise.

Of the eight planets, which has the shortest day?

Jupiter. Despite its massive size, the gas giant only takes about 10 hours to perform a full rotation. Because of this, there are about 10,400 Jovian days in a single orbit around the Sun, as Jupiter's orbit takes 11.86 Earth years.

What is the only day of the week named after a planet?

Saturday, after Saturn. Sunday and Monday are named after the Sun and Moon, while Tuesday, Wednesday, Thursday, and Friday get their names from Norse gods Tyr, Odin, Thor, and Frigga. This convention with Saturday comes directly from the Romans, who associated the day with the planet.

Which three neighboring planets spell the name of another celestial object with their first letters in order from Jupiter outward?

Saturn, Uranus, and Neptune spelling out "Sun." Although this is, of course, a complete coincidence, this wouldn't have happened if Uranus had received a name from the god's Roman equivalent, Caelus. Breaking away from the traditional naming pattern inadvertently created an acronym!

CHAPTER TWO

THE SOLAR SYSTEM

Do you capitalize "Solar System," or is it lowercase: "solar system"?

It depends. Since "solar system" can mean any system with one or more stars, saying "our solar system" is using it as a generic term. Saying "the Solar System" is like saying "the Moon" or "the Sun." Establishing a "the" in front of it, we use that as an official term, which is why it becomes capitalized.

What is the artificial object farthest out from the Sun?

The Space probe "Voyager 1." Anticipation had surfaced for its launch in 1977. It would become the first artificial object from Earth to reach interstellar space in 2012. Assuming it does not collide with anything, it will drift in interstellar Space for the rest of eternity.

What were gold-colored and included on both "Voyager 1" and "Voyager 2" as a message for any other intelligent life which might find it?

A phonograph record. On the outside is a map of directions to our Sun. The legend itself includes greetings in fifty-five languages and music from around the world. Astrophysicist Carl Sagan called the record a "bottle in the cosmic ocean."

What is escape velocity, and what does it have to do with the "Voyager" probes?

Escape velocity is how fast something needs to travel to resist a gravitational pull. Earth's escape velocity is about 7 miles per *second*! Although they're slowing down due to the Solar System's gravity, the probes are each traveling at about 10 miles per second, more than enough to escape the Solar System.

How many nuclear bombs would be needed to create the equivalent amount of energy that the Sun produces every second?

400 *billion*, and, yes, that does happen every *second*! The reason for this insane amount of energy is the famous equation, $E = mc^2$, which equates mass and energy with the speed of light. If we could harness just one-thousandth of that energy, we could satisfy our current energy needs for about 500 years!

True or false: there are no dwarf planets closer to the Sun than Neptune?

False! There is one dwarf planet, Ceres, which orbits within the asteroid belt. Ceres alone is close to 25% of all mass in the asteroid belt.

True or false: if you were to travel through the asteroid belt, you would likely collide with an asteroid?

False! Although the asteroid belt contains many asteroids, they're mostly small and spread far apart. Fortunately, this idea is just an invention of science fiction. Outer Space is called that for a reason - there's a lot of room out there!

What keeps planets, moons, and even asteroids in stable orbits?

Orbital resonance occurs when two or more objects orbit around the same mass, following a pattern, keeping them from knocking each other out of orbit. For instance, for every four rotations, Io makes around Jupiter, Ganymede makes one, and Europa makes two.

What takes up only 0.2% of the mass in our Solar System?

Everything but the Sun! (About one million planets the size of Earth could fit inside it!) That 0.2% includes the eight planets, a minimum of five dwarf planets, and thousands of asteroids. Oh, yes - there are probably a staggering three *trillion* comets and similar chunks of snow and ice hurtling around!

The Sun creates energy by combining hydrogen atoms to make helium, which releases massive amounts of energy. Do you know the name of this process?

Nuclear fusion! This process happens in the Sun's core, which is millions of degrees hot, but it takes *170,000 years* for that energy to reach the surface and then be sent outward in Space, including toward Earth.

How far along is the Sun in its lifespan?

About halfway. The Sun is about 4.6 billion years old, and estimates for its remaining lifetime range from 4.5 to 5.5 billion years. If the Sun were a human who lived 80 years, it would be between 36 and 40 years old right now, depending on how much time it has left.

When the Sun reaches the end of its lifespan, it will transform into what kind of star?

A Red Giant. By that time, about 5 billion years from now, the Sun will have expanded into a red giant. This expansion will likely vaporize Mercury, Venus, and Earth, as well as possibly Mars. This development will also move the zone of habitability much farther out into the Solar System, possibly as far as Neptune's moons or even Pluto.

True or false: planetary orbits around the Sun are circular?

False! While often depicted as circular, planetary orbits are more or less oval-shaped; the formal word is elliptical. All planets have elliptical orbits, although the orbit of Venus is closest to being truly circular.

Why are all the planets in our Solar System on the same orbital plane?

While there are slight variations in the angles, all planets are on the same orbital plane because that was the same plane as the disk that formed around our Sun. Momentum and gravity maintained this plane as planets formed.

Why do all the planets in our Solar System orbit in the same direction?

Because of angular momentum, when the Solar System formed, objects were orbiting in all directions. As celestial objects going one way collided with a body of mass going another, they canceled each other out and left the ones going in the most common direction.

True or false: will the Moon entirely block all light from the Sun in all solar eclipses?

False! In a total solar eclipse, this does happen, and the area in the path of a total eclipse will seem as if daytime has shifted to mimic night. However, in partial solar eclipses, a shifting part of the Sun is always visible during the Moon's transit.

Are solar eclipses unique to Earth?

No! The requirement for a partial eclipse is that a planet has a moon and the moon orbit takes it in front of the Sun. For a total eclipse, the moon has to be as large as, or larger than, the star in the planet's sky.

If solar eclipses happen when the Moon passes in front of the Sun, why doesn't one happen every month?

Orbital alignment. It takes about one month for the Moon to complete its orbit. However, the Moon's orbit is tilted slightly compared to the Earth's around the Sun, which is why it often misses. An eclipse can only happen when the planes of the two orbits align.

Do you know the official name of the icy-body belt beyond Neptune?

The Kuiper belt. All known dwarf planets, except for Ceres, exist in this belt. Because the belt begins after Neptune, every object found within it can be classified as a "trans-Neptunian object," defining the orbit of Neptune as the inner boundary of the Kuiper belt.

What is the most massive trans-Neptunian object?

While you might be inclined to say Pluto, the dwarf planet, Eris, has more volume. Its similar mass to Pluto inspired astronomers to rethink their classification as a planet, eventually leading to the definition change. (Eris is also about three times as far away from the Sun as Pluto!)

Do you know the nickname astronomers gave to Eris before it received its official name?

Xena! Inspired by the television series "Xena: Warrior Princess," and came about partly because the hypothetical tenth planet (Pluto was still considered number nine) was called "Planet X." Xena also started with the rarest letter in the English alphabet.

True or false: some astronomers believe there may be a ninth planet on the distant edges of the Solar System?

True! Although this is far from an agreement if Planet Nine is feasible, those who believe it could exist point to a clustering of some trans-Neptunian objects, which hints that a yet-to-be-discovered planet could be causing this through its gravity.

True or false: compared to the age of the Universe, the Solar System is young?

True! The Universe is approximately 14 billion years old, while the Solar System is only about 4 billion years old. To put this into perspective, if the Universe were 77 years old, our Solar System would only be 22!

Which was more recent: the extinction of the dinosaurs or the last time the Sun made an entire orbit around the galaxy?

Dinosaur extinction! Along with our Solar System, the Sun takes 250 *million* years to complete an orbit around the galaxy, while the dinosaurs only went extinct about 65 million years ago.

How fast, then, does the Sun, and therefore the Solar System, orbit around the galaxy?

The Solar System, pulled by the Sun, orbits the galactic center at a speed of 149 miles per second, nearly ten times escape velocity for the Solar System. In one hour, it covers a distance of 536,865 miles! Even at this speed, though, it would take 1,250 years to travel one light-year.

True or false: astronomers have definitively established the Solar System's outer edge?

False! One idea is that the Solar System finishes after Neptune's orbit. Still, another viewpoint is that this ends when the Sun's gravity is matched by that of other stars, which astronomers estimate could be anywhere from 0.8 light-years to 2 light-years away from the Sun.

Which one of the minor members is most likely the edge of the Solar System?

The Oort cloud region. Beyond the Kuiper belt. It is estimated to be two light-years away. This edge is what gave astronomers their best estimate of the Solar System's dimensions. The end of the Oort cloud signals the end of the Sun's gravitational dominance on objects.

Why is it called the Oort cloud and not the Oort belt?

Unlike the Kuiper belt, which exists on the same plane as the planets, the Oort cloud is spherical and extends in every direction from the Sun. "Voyager 1" will reach the cloud in about 300 years and could take up to 30,000 years to leave it.

How did the Oort cloud initially form?

It started from ejected pieces by planets, moons, and icy comets. After the planets formed, the leading idea is that they launched many of the leftover chunks out of the Solar System, where gravity from the galaxy and the Sun stabilized them into this region.

What is the closest star to the Solar System?

Proxima Centauri. Classified as a red dwarf and remains part of a triple-star system called Alpha Centauri. The star's distance is approximately 4.25 light-years away. Even "Voyager 1," traveling at just over 38,000 miles per hour, would take well over 70,000 years to reach the star.

Does everything in the Solar System have a gravitational effect on everything else?

Yes! Anything with mass has a gravitational effect. The invisible response is almost unnoticed due to the distance between objects and because the Sun holds 99 percent of our Solar System's mass.

Which small celestial bodies are just leftover material from the formation of our Solar System?

Asteroids! Most of them orbit between Jupiter and Mars.

Icy material makes up the majority of which small bodies?

Comets. They come from the Kuiper belt or beyond.

Which small bodies change names when they enter Earth's atmosphere?

Meteors. Thousands of them enter Earth's atmosphere every year.

Like the Moon, the Sun has an official name aside from simply "the Sun." Do you know the name?

"Sol," pronounced like "soul," and from where we get the term "solar." Although "Sol" comes from French, and thereby Latin, and "sun" comes from Old English, both have a common origin going back thousands of years to the same ancestor language for Latin and Old English.

It's neither a planet nor a moon, and these celestial objects share the same orbit as a planet or moon. Do you know what astronomers name them?

Trojan asteroids. As most of the early ones discovered, named after figures from the mythological Trojan War. Asteroids will maintain this same distance in the rotation if placed 60 degrees before or behind a planet or moon in its orbit. So, from the world's perspective, they don't move.

In which planet's orbit were the first trojan asteroids discovered?

Jupiter, with the first trojan discovered back in 1906! While Jupiter is most notable for them, Venus, Earth, Mars, Uranus, and Neptune also have at least one each. Although Saturn doesn't have any known trojans, its moon, Tethys, has two of them, one in front and another in back, in its orbit around Saturn.

How many trojans orbit Jupiter, then?

While there isn't an absolute number, some astronomers believe there could be as many, if not more, Trojan asteroids in Jupiter's orbit than there are in the *entire* asteroid belt!

CHAPTER THREE

BEYOND OUR SOLAR SYSTEM

What is a large cluster of stars, star systems, gas, and dust called?

A galaxy! On average, just one star cluster contains enough matter to equal about 100 *billion* suns! To put that into perspective, that's about one sun's worth of weight for every human that has ever lived on Earth.

Why is our galaxy called "the Milky Way"?

To early humans, it appeared as a milk-like substance spread out across the night sky. This milky appearance gave rise to the word "galaxy," from the Greek *galaxias,* meaning "milky." So, in a way, saying "Milky Way Galaxy" is like saying "Milky Way Milky"!

How far away from the Milky Way's galactic center is the Solar System?

Approximately 27,000 light-years. To put that into perspective, if you could travel at ten times the speed of light, you could fly around Earth at the equator about 75 times per *second.* If you left Earth traveling at that speed when Rome was organized in 753 BC., you'd have arrived at the galactic center just after the Second World War.

What is at the center of our galaxy, the Milky Way?

A supermassive black hole, with a mass estimated at around 4 million suns. The gravitational pull stretches across a region approximately 100,000 light-years in every direction, holding at least 100 billion stars and even more planets.

How long ago did astronomers realize that there were other galaxies in our Universe besides the Milky Way?

Remarkably, astronomers finally resolved this debate in the 1920s. Before then, astronomers had noticed nebulas that didn't seem to form stars, and the astronomers weren't sure whether these nebulas were in our galaxy or were beyond it. The realization that these objects were other galaxies caused a sea change in astronomy.

More than two-thirds of galaxies take the same shape as a nautilus shell, including the Milky Way. Can you name the shape?

A spiral. The familiar pattern is seen throughout nature ranging in size, from shells to hurricanes to galactic shapes. An elliptical form is the next most common shape for a galaxy; that means it's roughly circular, like our Solar System.

True or false: the ideal zone for planetary life, also known as the Goldilocks Zone, varies depending on the star system?

True! The Goldilocks Zone depends on three factors: type of star, distance from that luminary, and how much heat an atmosphere can trap and release. Variation in even one of these can change where the ideal zone would be.

What is a planet that doesn't orbit any star called?

A rogue planet! This malicious-sounding planetary mass would be a planet if it were within a solar system, though being starless technically doesn't fit the formal definition of one. A rogue planet wanders alone across a galaxy. Unbound by any gravity having a strong pull on it.

Could a rogue planet support life?

Not on its surface, which would be unimaginably cold (near absolute zero). However, deep within a heat-producing core, it possibly could, especially if it had an underground ocean like the one astronomers think exists on Europa, Jupiter's moon. Water is a crucial ingredient of life as we know it.

Star-bound planets Vs. Rogue planets, which are most likely to be more plentiful in our galaxy?

While there's currently no way to calculate this with precision, rogue planets are mathematically more likely to outnumber star-bound planets by 50%. So, for every two star-bound planetoids, there may be *three* rogue ones!

Could there be rogue planets outside of a galaxy?

Almost certainly not. Since rogue planets originate from galactic material, the galaxy has to come first, and therefore, its gravity would most likely keep the globe within the Universe. Even if one were outside a galaxy, the overwhelming gravity would eventually pull it into orbit.

How do rogue planets form?

Gravity or impact. One shared theory is that expelled rogue planets happen during a chaotic and powerful formation of their parent solar system. It could also be that the gravitational pull or an impact with another object kicks it out of its orbit around a fully-formed star.

What is a planet that orbits a star other than our Sun?

An exoplanet. The prefix "Exo" comes from Greek and refers to something that is "outside." Exoplanets are planets that exist outside of our Solar System; the prefix set out to specify that. Since the 1990s, thousands of exoplanets have amazed astronomers.

How do we find exoplanets?

If they're big, that helps spot them, particularly if they're close to their star; the star's gravity will have a measurable effect on their orbit. A later method for finding smaller exoplanets looks at how starlight is interrupted from a transiting planet and then uses that to determine its size.

What is the most massive known exoplanet?

GQ Lupi b. It's estimated to be three times the mass of Jupiter, which is 1,300 times the mass of Earth. GQ Lupi b's mass remains an estimate only due to the inexact methods currently available to find and study exoplanets from this distance.

What is the closest known exoplanet to Earth?

Proxima Centauri b, orbiting the red-dwarf star Proxima Centauri, in the Alpha Centauri trinary system. Astronomers estimate that this "super-Earth," discovered in 2016, has about 1.27 times the mass of Earth and takes only 11.2 days to orbit its star. One Earth year is equal to 32.6 Proxima-Centauri-b years.

What is the farthest known exoplanet from Earth?

The farthest confirmed exoplanet is a tie between two, SWEEPS-04 and SWEEPS-11. Although not in the same system, each one is 27,710 light-years from the Sun, about as far as the Solar System is from the Galactic Center.

Would life on other planets have to be based on Carbon?

No. Carbon is the building block of life because its atoms can attach to up to four molecules simultaneously. Silicon can also bond with up to four atoms and is relatively commonplace. However, silicon-based compounds are much rarer than Carbon-based ones, indicating Carbon-based life is more likely.

Can a planet's proportion be more significant than its star?

Yes. While humans have never found an instance where this is the case, it is mathematically possible. However, it's also doubtful for this to be the case with one like our Sun. The type of star is crucial to this being a mathematical possibility.

Visible to the naked eye, what is the closest region to Earth, forming stars and solar systems?

The Orion Nebula. Burning brightly inside the constellation of Orion, classified as a diffuse nebula. The nebula is 1,300 light-years away and 25 light-years across. The specific region where stars form is a "stellar nursery," but this is only one type of nebula.

How many nebula types are there?

Four: (1) star-forming, or emission, (2) planetary, formed when a star dies, (3) supernova remnant, where star material fragments spread far out, and (4) dark, which we can't directly see! We can only observe a dark nebula by what it's covering behind it.

Because of its shape and active star formation, which nebula has the nickname "The Pillars of Creation"?

The Eagle Nebula. Officially named "Messier 16," the nebula earned its nickname because of its pillar shape and its active star formation. Another part of it looks like an eagle with its wings prominently spread out.

Will our galaxy ever collide with anything?

Yes, but not violently. In 4 billion years, the Andromeda Galaxy will combine with ours due to the pull of gravity between them. There will be very few physical collisions, though, due to the vast emptiness of the galaxy.

How do we know that the Andromeda Galaxy is slowly moving toward us?

Andromeda's light is shifting blue, indicating it's moving toward us. The Doppler effect causes this. Every galaxy, except for Andromeda, emits light in the red direction, signaling that they're moving away from us.

What the heck is the Doppler effect?

Ever stood on a sidewalk and heard an ambulance drive past you? The sound is drastically different when the siren is moving toward you versus when it's moving away. Light is a wave, too, so this same shift happens with light frequencies from other galaxies since they are simultaneously moving.

This type of matter makes up 27% of our Universe, yet we can't see it. What name was it given?

Dark matter. Scientists aren't sure what it is exactly, but there wouldn't be enough mass to keep our galaxies bound together through gravity without it. When they ruled out other options, they called it "dark matter" because it didn't interact with light.

If a star is more than a billion light-years away, it is called what?

A ghost star. Because not only has the starlight taken so long to reach us, and therefore stars would currently be far away from where we see them as being, but the stars billions of light-years away no longer exist. We see what they looked like long ago.

Astronomers pinpointed the moment the Universe began expanding as this term?

The Big Bang. However, the term "Big Bang" implies a single, massive explosion, when the truth is, it was just the start of the expansion. The "Big Start" may be a more accurate title.

What is the primary focus of the Big Bang model?

Track the expansion backward in time. In the early 20th century, scientists realized that the Universe was expanding, so they realized that we could find the Universe's beginning if one traced that expansion back in time. In the genesis, everything stood concentrated at a single, dense, unimaginably hot point.

What evidence is there for the Big Bang model?

Microwave radiation. Astronomers have found this radiation coming from all directions in Space, and they've determined that this is light produced near the Universe's beginning. Because the Big Bang model predicted this precisely, the evidence confirmed that the model was correct.

How do astronomers know that the Universe is approximately 14 billion years old?

One method is to look at the most distant stars and figure out the age of the light reaching us, which gives us a minimum maturity of the Universe. However, astronomers honed in on a more accurate estimate using the microwave radiation all around the Universe left over from the Big Bang.

What has a gravitational pull so strong that not even light can escape it?

A black hole. The name can be misleading; they're not "holes," but extensive collections of matter, like planets and stars. Because they absorb all light, black holes can only be observed by how they affect the light around them, warping and disrupting it.

True or false: if the Sun were to change to a black hole with the same mass instantly, will all the planets be pulled into it?

False! What makes a black hole's pull so strong is the density of its matter, as opposed to simply its mass. While the loss of the Sun's heat would cause all planets to freeze, and that could, in turn, affect their rotation, they would maintain their orbits around their new host.

Is there anything that can travel faster than the speed of light?

Yes, from a certain point of view. There are galaxies beyond the edges of the observable Universe that we will never see because they are moving away from us faster than the speed of light, and so their light could never reach us.

How is it possible for those galaxies to travel past the speed of light?

Think about the Universe like a rubber ruler. If you stretched it to double its length, all distances would double. For us, it seems like things are moving faster than the speed of light, but it's Space itself that is expanding.

If everything is expanding, why haven't our planets spread any farther out?

Because of gravity, the amount of matter in solar systems and galaxies is significant enough and close-together enough to resist this expansion. Still, intergalactic Space is not, so it continues to expand.

What 19th-century poet was one of the first people on record to suggest the possibility of an expanding Universe?

Edgar Allan Poe attempted to explain why the night sky failed to fill with starlight overwhelmingly. The idea of a static Universe was accepted by Albert Einstein when he came up with his theory of relativity, though that theory later helped disprove the static-Universe hypothesis.

What are the two most common chemical elements in the Universe?

Hydrogen and helium. Almost 100% of all known matter is of one of these two. The death of supernovas caused all elements in the periodic table because their cores were dense enough to fuse hydrogen and helium atoms, except synthetically made elements by humans.

What is the last stage in the life cycle of a star?

A supernova. When a star goes supernova, it explodes; these are the giant explosions known in the Universe. The tremendous speed of the material causes it to spread far out into Space, and the material eventually becomes part of new systems.

Is it possible to see a supernova with just a telescope?

It is, but it's also doubtful. Sadly, most supernovas in the Milky Way likely are obscured by dust, and the last one viewed directly in our galaxy was by Johannes Kepler around 400 years ago. Supernovas in other galaxies are more visible but require *large* telescopes to see.

Where is the center of our Universe?

Nowhere! Because things are expanding everywhere, from our point of view, we're the center of the Universe, but there is no objective center. Everything spread out from a single point-in-time with the Big Bang, but that single point is now the entire Universe.

Do you know the name given to objects the size of a city, with more mass than the Sun, and which appear to flicker?

Pulsars! They aren't stars but vast collections of matter, not dense enough to be a black hole but has enough density to emit radio waves. Pulsars seem to flicker because they spin and are only visible when facing us, like the beam from a lighthouse.

CHAPTER FOUR

ALL ABOUT ASTRONOMY

What exactly is astronomy in the first place?

Astronomy is the study of objects and phenomena that occur outside of Earth and its atmosphere. Points of interest include measuring distances to planets, stars, and other galaxies, looking for life on other planets, and studying how the Universe works.

Where and when was the first known telescope invented?

In the Netherlands, in 1608, a lens-maker applied for a patent on something that fits the definition of a telescope. Although the application never completed approval, and other similar ideas in development elsewhere, this is the first time that a proper telescope appears in historical records.

Who was the first person known to make astronomical observations using a telescope successfully?

Galileo! Not only did he have the first up-close views of the five known planets (including being the first to see Saturn's rings) and Jupiter's four biggest moons, but his use of the telescope also helped him realize that Earth orbits the Sun.

What made Galileo's telescopic observations so effective?

While he didn't invent the telescope, Galileo used his engineering skills to improve the models' design. So, it wasn't just how he used it. These modifications ultimately gave Galileo an advantage in making astronomical observations.

True or false: Space-based telescopes have a better view of the stars than ground-based ones?

True! However, this isn't because they're closer to the stars. Earth's atmosphere distorts and dims light (this is why stars appear to twinkle), so while modern ground-based telescopes can view a lot more than historical assemblies, Space-based ones will always have a more transparent and better view.

The Large Binocular Telescope is the largest in the world in terms of its lens-opening size. In which U.S. state is the LBT located?

Arizona. Telescopes need to be as far away from artificial light as possible because the light produced by humans at night blocks out starlight. Even though it's on a mountain far away from cities, light pollution still sometimes gets into its images.

What is the James Webb Space Telescope?

A decades-long collaboration between NASA, the European Space Agency, and the Canadian Space Agency, Webb is the largest Space telescope ever assembled. It will allow astronomers to study every stage of the Universe's lifespan, as well as the formation of solar systems.

Who was the first person to calculate the circumference of the Earth?

Eratosthenes, born in Cyrene (modern Libya,) is primarily known as a mathematician. Accurately logging the Earth's size allowed later astronomers to calculate distances to both the Moon and Sun, as well as their sizes. His measurement led to humanity's ability to measure cosmic distances.

True or false: before Christopher Columbus sailed, most people believed the Earth was flat?

False! There is no historical basis for this idea. Ancient peoples knew the Earth was round. They could see that ships sailing away from the coast would dip under the horizon. They also noticed the shadow Earth casts on the Moon in a lunar eclipse.

Wait, is our planet called "Earth" or "the Earth"?

Both! It depends on who you ask and how they're using it. Since we live on it, Earth feels special, hence "the" in front of it. However, unlike with the Moon, we don't call other planets "earths," so "Earth" by itself isn't confusing.

Who was the first astronomer to realize that planetary orbits were elliptical rather than perfectly circular?

Johannes Kepler. Before him, Nicolaus Copernicus depicted orbits as circular, but Kepler realized that this model did not match the data collected. He reasoned that the Sun was not at the exact center of the orbits.

What is an astronomical unit?

It's the approximate distance between Earth and Sun. Since 2012, astrophysicists agree on exactly 92,955,807.273 miles. Knowing the astronomical unit allowed astronomers to calculate distances to the other planets and, eventually, to nearby stars.

Aside from the last name of a Pixar character, what is a light-year?

A light-year is the amount of distance that light can travel in an Earth year. The vastness of Space requires a large unit of distance to measure it. Also, light moves at a constant rate - 67 million miles per hour - making it a reliable gauge.

How does a light-year compare to an astronomical unit?

Well, it takes light about 8 minutes to reach Earth, so one astronomical unit is approximately equal to 8 light minutes. One light-*year* is equivalent to 63,241.08 astronomical units exactly, based on the official definition of an astronomical unit.

What was the first star to have its distance from the Solar System accurately measured?

61 Cygni, measured in the year 1838. However, because it was a massive number of astronomical units away (720,000), astronomers developed the light-year as a more convenient way of measuring distance. With this unit, the star was merely 11.4 light-years away.

What unit does Han Solo incorrectly use to measure time in "Star Wars Episode IV: A New Hope"?

A parsec. Equal to about 3.26 light-years. Like a light-year, a parsec may sound like a measurement of time, but it measures distance. Astronomers prefer to use parsecs over light-years for their work, but in announcements to the public, they often use light-years since that unit is more commonly known.

True or false: Isaac Newton's theory of gravity is the most accurate one that exists today?

False! While Newton's ideas on gravity hold up with our experiences, Albert Einstein's theory of gravity replaced it in the early 20th century because it was more accurate.

What made Einstein's theory of gravity more accurate than Newton's?

Newton described gravity as an attractive force between two bodies. While that *seems* reasonable, what Einstein realized is that mass causes attraction indirectly rather than directly. Einstein's ideas could explain phenomena that Newton's couldn't, so it won out.

True or false: a theory eventually becomes law once enough evidence comes in to support its validity?

False! To use gravity as an example, the "law of gravity" states that objects are attracted to the center of mass. On the other hand, a "theory of gravity" attempts to explain *why* this happens in the first place. One theory can be better than another, but both take the law to be true.

What is the cost for one astronaut's spacesuit in American dollars?

$12 million! Each one is so expensive because a suit has to be a self-contained livable habitat for the astronaut and can withstand everything in Space, from solar radiation to the vacuum itself.

True or false: objects in low-Earth orbit (the closest possible orbital region to Earth) stabilize here because they are merely floating in Space?

False! Objects in low-Earth orbit steady there because the speed they are falling toward Earth is slower than the speed compared to Earth's rotation, so one could say gadgets are "falling with style."

What is a room with two doors that allows astronauts to safely enter and exit a spacecraft to keep from damaging the living areas?

An airlock. With only one door, everything inside would be sucked into Space as soon as it opened. With two doors, the astronaut can be in the central room while one door opens and the other remains closed, keeping everyone safe.

How many astronauts are on the International Space Station at one time?

Six. Since beginning operations in 2000, 242 people from 19 countries have been on the Space Station, upholding the "international" ideal that brought the station into being. It can take as little as *four hours* from launch for a craft to arrive at the ISS!

True or false: the International Space Station is visible from Earth with the naked eye?

True! Because the ISS has more than an acre of solar panels. That makes it the third-brightest object in the sky at certain times of the day. Even if you live in a big city, you could see it passing overhead at dawn or dusk, depending on where you are. You can learn more at spotthestation.nasa.gov.

True or false: there is no gravity on the International Space Station?

False! Although ISS inhabitants float the same way they would if they were in deep Space, the reason for their floating is because they're "falling with style." Just as when anything is falling, they're weightless, but because they never reach the ground, they remain in permanent freefall, which makes them float.

The radio signal sent from Earth to an artificial satellite is called what?

An uplink. In this case, "satellite" refers to something that orbits Earth, which would include spacecraft, space stations, and what we traditionally think of as a satellite. A signal sent the opposite way is a "downlink."

Can you name the signature fuel of the American Space program?

Liquid hydrogen. Every ounce matters for a space mission, and since hydrogen is the lightest element, engineers can pack more of it into a rocket than any other fuel. It remarkably has to be stored in between -434.6 and -423 °F to be in liquid form!

What was so special about the spacecraft "Mariner 4," launched November 28, 1964?

It was the first spacecraft to perform a flyby of another planet! The craft reached Mars in July 1965 and transmitted images of the Red Planet back to Earth. Its images of a dead, dry sphere radically changed astronomers' speculations about the possibility of Martian life.

On April 8, 2021, NASA's Martian helicopter, "Ingenuity," took the first-ever flight on another planet. What was so special about the swatch of muslin it carried?

The material, taken from the plane's left wing, that the Wright brothers used for their first powered flights in 1903! So, this was a sort of commemoration. Thoughtfully, the swatch will forever tie the start of aviation on Earth to the beginning of aviation outside of Earth!

What is the term used to refer to astronauts from the Soviet Union and now Russia?

"Cosmonaut," from the Greek word *cosmos*, referring to the Universe. *Astro* also originates from a Greek word, referring to "star," which can be seen in the famous Latin phrase "ad astra," meaning "to the stars." (Latin borrowed the word from Greek, as well.)

19th-century author Jules Verne came up with the idea of a gun as a spaceship-launching device. How did pioneering physicist Isaac Newton likely inspire Verne's vision for Space travel?

Newton conducted a thought experiment late in his life, called Newton's Cannonball, where he imagined a cannonball being fired at such a high speed that it wouldn't fall back to Earth. The aforementioned is what happens with objects in orbit, and it looks toward developments about three centuries later than Newton's lifetime.

Which former country was the first to send a man into outer Space?

The Soviet Union. In 1961, it sent Yuri Gagarin far enough upward that his ship was in orbit around Earth. That accelerated the Space Race, beginning in the late 1950s between the Soviet Union and the United States. However, the United States became the first to land men on the Moon.

Who was the first American to successfully launch and be in Outer Space?

Alan Shepard, only a month after Gagarin. though, unlike Gagarin, Shepard was never in orbit around Earth. (The first American in orbit was future U.S. Senator John Glenn.) However, in 1971 Shepard became the fifth person to walk on the Moon!

True or false: the famous Neil Armstrong quote, "One small step for man, one giant leap for mankind," is a misquote?

True! According to Armstrong, he said "one small step for *a* man," but his communication cut out at that moment. Without "a" the quote doesn't make sense, as "man" and "mankind" would mean the same thing.

Why do the footprints of astronauts remain on the Moon for a long time, possibly forever?

Luna lacks a troposphere similar to Earth. The atmosphere sweeps away footprints, over time, due to wind and water. That doesn't exist on the surface of the Moon. Volcanic activity also doesn't happen there. A meteor impact would destroy a footprint, but the chances of a hit being that specific are low.

Although Sally Ride was the first American woman in Space, the Soviets sent a woman up earlier than her. Do you know her name?

Valentina Tereshkova. Although Tereshkova launched into Space on June 16, 1963, Sally Ride wouldn't become the American trailblazer until 20 years later, on June 18, 1983. However, Ride's selection was announced back in 1978, attesting to the years-long preparation required for astronaut missions.

In what year was the word "spaceship" first used in the form we use it today?

1894! John Jacob Astor was not the first person on record to use that term, but he is the first one known to use it to mean a "spacecraft." He did this in his science fiction novel, "A Journey in Other Worlds."

What was the first film to depict human travel to another celestial body?

The 1903 French short film "A Trip to the Moon." Mainly, though not exclusively, inspired by works of science-fiction writer Jules Verne. Less than 60 years later, Space would welcome the first man launched in real life.

Which country has sent the most astronauts to Outer Space?

The United States, with 339, more than 61% of the total number of people who have been to Space. In second place is Russia, with 117 (if one includes Soviet-era cosmonauts), and in third is Japan, with 12.

Which three countries have launched crewed missions to Outer Space?

The United States, the Soviet Union/Russia, and China. However, about 40 countries have had their citizens sent to Space when one looks at crew members, though the missions themselves fell under one of those three banners.

True or false: it's difficult to pin down exactly where a planet ends and Outer Space begins?

True! There is no exact method for figuring out where the line should be drawn, even among astronomers. One suggestion for Earth is called the "Karmin Line," named after the astronomer who proposed it, which marks where aeronautic machines such as airplanes would no longer function.

What is Space junk?

Artificial debris in orbit around Earth, most of it coming from Space missions. While the lion's share is somewhere between half-an-inch and 4 inches in size, because these objects travel at up to 4 miles per *second*, collisions between them and functioning equipment can pose a risk to Space missions.

Why is something twice as far away from another object only 25% as bright as that object, assuming they produce the same amount of light?

The scientific law calls this the inverse-square law. Imagine shining a square light on a wall from three feet away, then six. The light is the same, but the area covered is twice as big, calculating the brightness will go down by ½ on each side. One-half squared is ¼ or 25%.

Why is the inverse-square law so crucial for astronomers?

If astronomers know what kind of star they're looking at, they know how bright it should be. Looking at how bright it appears to us, they can work backward to figure out how far away the star positions itself, to appear as dim as it does.

What does the acronym "SETI" stand for in astronomy?

"Search for Extra-Terrestrial Life," or, in other words, looking for aliens. Though microscopic life may exist outside of Earth within the Solar System, astronomers have ruled out intelligent life. The search for martian life switched focus onto exoplanets with habitability like Earth.

Has there ever been any possible alien communication with Earth?

A few detected radio signals fit what alien civilizations might produce, including one in 2019 from our nearest neighbor star, Proxima Centauri, and another in 1977 known as the WOW signal. However, none of these signals have ever been re-detected, making them unlikely to be from intelligent alien life.

Could there be intelligent life out there?

Just because we haven't found any so far does not mean that there isn't any. The Universe is unimaginably vast, and we only occupy a tiny part of it. With billions of galaxies and billions of stars in each one, the chances seem pretty good.

FINAL WORDS

Thank you so much for coming on this journey in *Fun Facts Space Trivia*! Science is seldom about the actions of individual geniuses but more to do with the collaborations of minds across both Space and time. You don't have to work in science to be a part of that conversation. Many citizen scientists have helped NASA with thousands of discoveries. However, even if you just enjoy gazing at the night sky, that means you're still part of that conversation.

We all started with curiosity that would eventually bloom into a passion.

RESOURCES

Chapter One: Our Planets and Moons

- Barnett, Amanda, et al. (Updated **December 19, 2019**). *What is a Planet?* Nasa Science: Space System Exploration. **https://solarsystem.nasa.gov/planets/in-depth/**

- Barnett, Amanda, et al. *What is a Moon?* Nasa Science: Space System Exploration. https://solarsystem.nasa.gov/moons/overview/

- Barnett, Amanda, et al. *Pluto.* (Updated February 15, 2021). Nasa Science: Solar System Exploration. https://solarsystem.nasa.gov/planets/dwarf-planets/pluto/overview/

- Anissimov, Michael. (February 17, 2021). *What is the largest asteroid ever to hit Earth?* Info Bloom.com. https://www.infobloom.com/what-is-the-largest-asteroid-ever-to-hit-earth.htm

- Nasa.gov. "Archive of astronomy questions." *Why don't Mercury and Venus have moons?* https://image.gsfc.nasa.gov/poetry/venus/q328.html

- Barnett, Amanda, et al. *Ganymede.* Nasa Science: Solar System Exploration. https://solarsystem.nasa.gov/moons/jupiter-moons/ganymede/by-the-numbers/ and

- Barnett, Amanda, et al. *Mercury.* Nasa Science: Solar System Exploration. https://solarsystem.nasa.gov/planets/mercury/by-the-numbers/

- Barnett, Amanda, et al. (Updated June 27, 2019). *Titan.* Nasa Science: Solar System Exploration,. https://solarsystem.nasa.gov/moons/saturn-moons/titan/overview/

- Barnett, Amanda, et al. (May 1, 2019). *What does a sunrise/sunset look like on Mars?* Nasa Science: Solar System Exploration. https://solarsystem.nasa.gov/news/925/what-does-a-sunrise-sunset-look-like-on-mars/

- Barnett, Amanda, et al. (December 19, 2019) *Neptune* and *Uranus.* Nasa Science: Solar System Exploration. https://solarsystem.nasa.gov/planets/neptune/in-depth/ and https://solarsystem.nasa.gov/planets/uranus/in-depth/

- Barnett, Amanda, et al. (May 1, 2019). *What does a sunrise/sunset look like on Mars?* Nasa Science: Solar System Exploration. https://solarsystem.nasa.gov/news/925/what-does-a-sunrise-sunset-look-like-on-mars/

- Crash Course Astronomy. "Saturn #18." YouTube video, 12.15. May 22, 2015. https://www.youtube.com/watch?v=E8GNde5nCSg.

- Kikrop, Victor. (January 8, 2019). *Which Planets have Rings?* World Atlas.com. https://www.worldatlas.com/articles/which-planets-have-rings.html

- Barnett, Amanda, et al. (December 19, 2019). *Uranus.* Nasa Science: Solar System Exploration. https://solarsystem.nasa.gov/planets/uranus/in-depth/

- ibid. *Uranus* and *Neptune.* https://solarsystem.nasa.gov/planets/uranus/in-depth/ and https://solarsystem.nasa.gov/planets/neptune/in-depth/

- Brennan, Pat, et al. (March 22, 2021) *Gas Giant.* Nasa: Exoplanet Exploration. https://exoplanets.nasa.gov/what-is-an-exoplanet/planet-types/gas-giant/

- Brennan, Pat, et al. (March 22, 2021) *Terrestrial.* Nasa: Exoplanet Exploration. https://exoplanets.nasa.gov/what-is-an-exoplanet/planet-types/terrestrial/

- Barnett, *Uranus.* https://solarsystem.nasa.gov/planets/uranus/in-depth/

- Ormondi, Sharon. (July 26, 2019). *Are there volcanoes on other planets?* World Atlas.com. https://www.worldatlas.com/articles/are-there-volcanoes-on-other-planets.html

- Barnett, Amanda, et al. (February 15, 2021). *Jupiter.* Nasa Science: Solar System Exploration. https://solarsystem.nasa.gov/planets/jupiter/overview/

- Barnett, Amanda, et al. (December 19, 2019). *Saturn.* Nasa Science: Solar System Exploration. https://solarsystem.nasa.gov/planets/saturn/in-depth/ and Barnett, *Jupiter,* https://solarsystem.nasa.gov/planets/jupiter/in-depth/

- Nasa Science: Exploration Program, *Mars Facts*. Nasa.gov. https://mars.nasa.gov/all-about-mars/facts/

- Barnett, Amanda, et al. (December 9, 2019). *Uranus Moons*. Nasa Science: Solar System Exploration. https://solarsystem.nasa.gov/moons/uranus-moons/overview/

- Barnett, Amanda, et al. (December 9, 2019). *Mars Moons*. Nasa Science: Solar System Exploration. https://solarsystem.nasa.gov/moons/mars-moons/in-depth/#otp_how_mars_moons_got_their_names

- National Hurricane Center, Miami, Florida. *Saffir Simpson Hurricane Wind Scale*. https://www.nhc.noaa.gov/aboutsshws.php

- Candanosa, Roberto. (August 7, 2017). *Jupiter's Great Red Spot*. Nasa, Goddard Space Flight Center. Nasa.com. https://www.nasa.gov/feature/goddard/jupiter-s-great-red-spot-a-swirling-mystery

- Ryba, Jeanne. (November 22, 2007). *The Great Dark Spot*. Nasa https://www.nasa.gov/missions/solarsystem/grt_darkspot.html

- Barnett, Amanda, et al. (December 9, 2019). *Saturn Moons*. Nasa Science: Solar System Exploration. https://solarsystem.nasa.gov/moons/saturn-moons/in-depth/

- Educationaltechs.com. (July 10, 2020). https://www.educationaltechs.com/2020/06/moon-is-drifting-away-from-earth.html

- Doyle, Heather, and Kristen Erickson. (June 3, 2021). *How far away is the Moon?* Nasa Science; Space Place. https://spaceplace.nasa.gov/moon-distance/en/

- Bryner, Michelle. (November 14, 2012). *How did Earth get its name?* Live Science.com. https://www.livescience.com/32274-how-did-earth-get-its-name.html

- Koczor, Ron, and Tony Phillips. (October 7, 2003). *The Goldilocks Zone.* Nasa News. https://www.nasa.gov/vision/earth/livingthings/microbes_goldilocks.html

- SciShow Space. "Moxie and SpaceX Launch." YouTube video, 5.40. April 30, 2021. https://www.youtube.com/watch?v=wR_QgYyzwwQ,

- Fox, Karen. (February 3, 2021). *Earth's Magnetosphere.* Nasa.gov. https://www.nasa.gov/magnetosphere

- Barnett, Amanda, et al. (December 19, 2019). *Europa.* Nasa Science: Solar System Exploration. https://solarsystem.nasa.gov/moons/jupiter-moons/europa/in-depth/,

- Barnett, Amanda, et al. (June 27, 2019). *Titan.* Nasa Science: Solar System Exploration. https://solarsystem.nasa.gov/moons/saturn-moons/titan/overview/

- Barnett, Amanda, et al. (May 19, 2021). *Mars.* Nasa Science: Solar System Exploration. https://solarsystem.nasa.gov/planets/mars/in-depth/

- Barnett, Amanda, et al. (December 19, 2019). *Callisto.* Nasa Science: Solar System Exploration. https://solarsystem.nasa.gov/moons/jupiter-moons/callisto/in-depth/

- Phys.Org. (October 11, 2013). *An explanation of the rotational state of Mercury.* University of Namur. https://phys.org/news/2013-10-explanation-rotational-state-mercury.html

- Barnett, Amanda, at al. (December 19, 2019). *Jupiter.* Nasa Science: Solar System Exploration. https://solarsystem.nasa.gov/planets/jupiter/in-depth/

- Shupla, Christine, et al. (2021). *Shaping the Planets; Impact Cratering.* LPI Education. https://www.lpi.usra.edu/education/explore/shaping_the_planets/impact-cratering/

- Chu, Jennifer. (December 14, 2020). *Astronomers may have found a signature of life on Venus.* Massachusetts Institute of Technology. https://news.mit.edu/2020/life-venus-phosphine-0914

- Barnett, Amanda, et al. (December 19, 2019). *Neptune.* Nasa Science: Solar System Exploration. https://solarsystem.nasa.gov/planets/neptune/in-depth/

- Miriam-Webster.com dictionary. "Jovian." Accessed June 6, 2021. https://www.merriam-webster.com/dictionary/Jovian

- Barnett, Amanda, et al. (February 15, 2021). *Saturn.* Nasa Science: Solar System Exploration. https://solarsystem.nasa.gov/planets/saturn/overview/

- Barnett, Amanda, et al. (December 19, 2019). *Jupiter.* Nasa Science: Solar System Exploration. https://solarsystem.nasa.gov/planets/jupiter/in-depth/

Chapter Two: The Solar System

- Barnett, Amanda, et al. (February 17, 2021). *What spacecraft are headed into interstellar Space?* Nasa Science: Solar System Exploration. https://solarsystem.nasa.gov/solar-system/our-solar-system/overview/#otp_faq:_what_spacecraft_are_headed_into_interstellar_

- Espinoza, Luis, and Anil Natha. *The Golden Record.* Jet Propulsion Laboratory, California Institute of Technology. https://voyager.jpl.nasa.gov/golden-record/

- Canright, Shelley. (April 10, 2009). *Escape Velocity.* Nasa.gov. https://www.nasa.gov/audience/foreducators/k-4/features/F_Escape_Velocity.html

- Espinoza, Luis, and Anil Natha. *Where are the Voyagers now?* Jet Propulsion Laboratory, California Institute of Technology. Nasa.gov. https://voyager.jpl.nasa.gov/mission/status/#where_are_they_now

- Crash Course Astronomy. "The Sun." YouTube video, 12.03. March 19, 2015. https://www.youtube.com/watch?v=b22HKFMIfWo

- Dr Knowledge. (2006). *How much energy does the Sun produce?* Boston.com news. https://archive.boston.com/news/science/articles/2005/09/05/how_much_energy_does_the_sun_produce/

- Barnett, Amanda, et al. (February 15, 2021). *Ceres.* Nasa Science; Solar System Exploration. https://solarsystem.nasa.gov/planets/dwarf-planets/ceres/overview/

- SciShowSpace. "The most stable neighborhoods in the Universe." YouTube video, 6.11. March 26, 2019. https://www.youtube.com/watch?v=JxISBhMBDD4

- Barnett, Amanda, et al. (February 15, 2015). *The Sun.* Nasa Science; Solar System Exploration. https://solarsystem.nasa.gov/solar-system/sun/overview/

- Barnett, Amanda, et al. (December 19, 2019). *Our Sun.* Nasa Science; Solar System Exploration. https://solarsystem.nasa.gov/solar-system/sun/in-depth/

- Williams, Matt. (December 22, 2015). *What is the Life Cycle of the Sun?* Universe Today.com. https://www.universetoday.com/18847/life-of-the-sun/

- Betz, Eric. (February 6, 2020). *Here's what happens to the Solar System when the Sun dies.* Discover.com. https://www.discovermagazine.com/the-sciences/heres-what-happens-to-the-solar-system-when-the-sun-dies

- Barnett, Amanda, et al. (December 19, 2019). *Venus.* Nasa Science; Solar System Exploration. https://solarsystem.nasa.gov/planets/venus/in-depth/#otp_orbit_and_rotation

- Astrum. "The Real Reason all the Planets are on the same Orbital Plane." YouTube video, 8.23. March 22, 2020. https://www.youtube.com/watch?v=ceFl7NlpykQ

- Barnett, Amanda, et al. (December 19, 2019). *Venus*. Nasa Science: Solar System Exploration. https://solarsystem.nasa.gov/planets/venus/in-depth/#otp_orbit_and_rotation

- Geggel, Laura. (August 5, 2017). *Do Other Planets have Solar Eclipses?* Live Science.com. https://www.livescience.com/60037-do-other-planets-have-solar-eclipses.html

- Young, Alex. *How Eclipses Work*. Total Eclipse: August 21, 2017. Nasa.gov. https://eclipse2017.nasa.gov/how-eclipses-work

- Barnett, Amanda, et al. (December 19, 2019). *Eris*. Nasa Science: Solar System Exploration. https://solarsystem.nasa.gov/planets/dwarf-planets/eris/in-depth/

- Barnett, Amanda, et al. (December 19, 2019). *Hypothetical Planet X*. Nasa Science: Solar System Exploration. https://solarsystem.nasa.gov/planets/hypothetical-planet-x/in-depth/

- Matthews, Robert. *How long does it take the Sun to orbit the galaxy?* BBC Science Focus.com. https://www.sciencefocus.com/space/how-long-does-it-take-the-sun-to-orbit-the-galaxy/

- Williams, Matt. (February 14, 2017). *Distance and speed of Sun's orbit around galactic centre measured*. Universe Today.com. https://www.universetoday.com/133414/distance-speed-suns-orbit-around-galactic-centre-measured/

- Barnett, Amanda, et al. (March 18, 2019). *Where is the edge of the Solar System?* Nasa Science: Solar System Exploration. https://solarsystem.nasa.gov/resources/2232/where-is-the-edge-of-the-solar-system/

- Barnett, Amanda, et al. (December 19, 2019). *Oort Cloud.* Nasa Science: Solar System Exploration. https://solarsystem.nasa.gov/solar-system/oort-cloud/in-depth/

- Hille, Karl. (November 1, 2013). *Hubble's new shot of Proxima Centauri.* Nasa.gov. https://www.nasa.gov/content/goddard/hubbles-new-shot-of-proxima-centauri-our-nearest-neighbor/

- Espinoza, Luis, and Anil Natha. *Where are the Voyagers now?* Jet Propulsion Laboratory, California Institute of Technology. Nasa.gov. https://voyager.jpl.nasa.gov/mission/status/#where_are_th ey_now

- Childers, Tim. (September 4, 2019). *What's the difference between asteroids, comets, and meteors?* Live Science.com. https://www.livescience.com/difference-between-asteroids-comets-and-meteors.html

- Online Etymology Dictionary. "Sol" Accessed June 6, 2021. https://www.etymonline.com/word/sol

- Ibid. "Sawel" https://www.etymonline.com/word/*sawel-

- Turner, David. (February 22, 2021). *How were the Trojan asteroids discovered and named?* Nasa.gov. https://www.nasa.gov/feature/goddard/2021/how-were-the-trojan-asteroids-discovered-and-named, https://www.thoughtco.com/trojan-asteroids-3072197

- Hamilton, Calvin, J. (1995-2011). *Saturn's Trojan Moon, Telesco.* Views of the Solar System. Solarviews.com. https://solarviews.com/eng/telesto.htm

- Millis, John, P. (July 10, 2019). *Trojan Asteroids: What are they?* Thoughtco.com. https://www.thoughtco.com/trojan-asteroids-3072197

Chapter Three: Beyond Our Solar System

- Wordnik. "Galaxy: definitions." Accessed June 6, 2021. https://www.wordnik.com/words/galaxy

- Nagaraja, Mamta. (June 6, 2021). *Galaxies.* Nasa Science. Nasa.gov. https://science.nasa.gov/astrophysics/focus-areas/what-are-galaxies

- Online Etymology Dictionary. "Galaxy." Accessed June 6, 2021. https://www.etymonline.com/word/galaxy

- Mattson, Barbara, et al. (October 22, 2020). *Imagine the Universe.* Nasa: Goddard Space Flight Center. Nasa.gov. https://imagine.gsfc.nasa.gov/features/cosmic/milkyway_info.html

- Mohon, Lee. (July 23, 2019). *Galactic Center.* Nasa.gov. https://www.nasa.gov/mission_pages/chandra/images/galactic-center.html

- Kurzgesagt. "True Limits of Humanity." YouTube video, 11.40. May 11, 2021. https://www.youtube.com/watch?v=uzkD5SeuwzM

- SciShow. "How do we measure the distance of stars?" YouTube video, 9.51. September 8, 2014. https://www.youtube.com/watch?v=kyuI4n5ILP8, 7:28–

- Nagaraja, Mamta. (June 6, 2021). *Galaxies.* Nasa Science. Nasa.gov. https://science.nasa.gov/astrophysics/focus-areas/what-are-galaxies

- SciShowSpace. "Rogue Planets: Loners of the Universe." YouTube video, 3.20. July 15, 2014. https://www.youtube.com/watch?v=pAsACBDi_sk

- Otap, Lenka. (December 9, 2019). *Rogue Planets; the Lonely Nomads of the Universe.* Predict. Medium.com. https://medium.com/predict/rogue-planets-the-lonely-nomades-of-the-universe-3a6aa2b2bfb3

- Brennan, Pat. (Updated April 2, 2021). *What is an Exoplanet?* Nasa; Exoplanet Exploration. https://exoplanets.nasa.gov/what-is-an-exoplanet/overview/

- Doyle, Heather, and Kristen Erickson. (June 3, 2021). *What is an Exoplanet?* Nasa Science: Space Place. https://spaceplace.nasa.gov/all-about-exoplanets/en/

- The Nine Planets. *How Big is Jupiter?* https://nineplanets.org/questions/how-big-is-jupiter/

- Brennan, Pat. *Proxima Centauri b.* Nasa, Exoplanet Exploration. https://exoplanets.nasa.gov/exoplanet-catalog/7167/proxima-centauri-b/

- PHL. (December 5, 2015). *Top Ten Exoplanets.* Planetary Habitability Laboratory, University of Puerto Rico at Arecibo. http://phl.upr.edu/projects/habitable-exoplanets-catalog/top10

- Choi, Charles. (April 19, 2017). *Silicon-Based Life may be more than just Science Fiction.* Mach: Science. NBCnews.com. https://www.nbcnews.com/mach/science/silicon-based-life-may-be-more-just-science-fiction-n748266

- Antonellis, Jesse, and Chris Impey. *Silicon versus Carbon.* Teach Astronomy.com. https://www.teachastronomy.com/textbook/Life-On-Earth/Silicon-versus-Carbon/

- Plait, Phil. (February 17, 2021). *Can a Planet be Bigger than its Star?* Bad Astronomy. https://www.syfy.com/syfywire/can-a-planet-be-bigger-than-its-star

- Jones, Trevor. *What is a Nebula?* Astro Backyard.com. https://astrobackyard.com/what-is-a-nebula/

- Garner, Rob (editor). (October 22, 2019). *Messier 16: The Eagle Nebula.* Nasa, Hubble's Messier Catalog. Nasa.gov. https://www.nasa.gov/feature/goddard/2017/messier-16-the-eagle-nebula

- Crash Course. "Dark Energy: Cosmology Part 2, #43" YouTube video, 11.22. December 17, 2015. https://www.youtube.com/watch?v=gzLM6ltw3l0

- Crash Course. "Dark Matter: #41" YouTube video, 11.59. December 3, 2015. ://www.youtube.com/watch?v=9W3RsaWuCuE.

- Doyle, Heather, and Kristen Erickson. (June 3, 2021). *What is the Big Bang?* Nasa Science: Space Place. https://spaceplace.nasa.gov/big-bang/en/

- Ibid. https://spaceplace.nasa.gov/big-bang/en/

- Crash Course. "The Big Bang, #42." YouTube video, 13.22. December 10, 2015. https://www.youtube.com/watch?v=9B7Ix2VQEGo, 7:37–8:29

- SciShowSpace. "How do we know the age of the Universe?" YouTube video, 5.42. February 20, 2018. https://www.youtube.com/watch?v=tCn96DbBnB4.

- Garner. Rob. (Updated November 23, 2020). *What are Black Holes?* Nasa.gov. https://www.nasa.gov/vision/universe/starsgalaxies/black_h ole_description.html

- SciShow. "Climate change moved the North Pole." YouTube video, 4.53. May 7, 2021. https://www.youtube.com/watch?v=O_ui5uhLcyo

- Crash Course. "The Big Bang, #42." YouTube video, 13.22. December 10, 2015. https://www.youtube.com/watch?v=9B7Ix2VQEGo

- Crash Course. "Dark Energy, Cosmology part 2, #43." YouTube video, 11.22. December 17, 2015. https://www.youtube.com/watch?v=gzLM6ltw3l0

- Crash Course. "The Big Bang, #42." YouTube video, 13.22. December 10, 2015. https://www.youtube.com/watch?v=9B7Ix2VQEGo, 2:45–3:07

- May, Sandra, editor. (Updated July 16, 2018). *What is a Supernova?* Nasa.gov. https://www.nasa.gov/audience/forstudents/5-8/features/nasa-knows/what-is-a-supernova.html

- Cofield, Calla. (April 22, 2016). *What are Pulsars?* Space.com. https://www.space.com/32661-pulsars.html

Chapter Four: All About Astronomy

- Miriam-Webster.com dictionary. "Astronomy." Accessed June 7, 2021. https://www.merriam-webster.com/dictionary/astronomy

- "Telescopes." Www. bo. astro. it/dip.

- http://www.bo.astro.it/dip/Museum/english/can_int.html

- Crash Course. "Introduction to Astronomy #1." YouTube video, 12.11. January 15, 2015. https://www.youtube.com/watch?v=0rHUDWjR5gg

- Green, Richard, John Hill, and James Slagle. "The Large Binocular Telescope." University of Arizona, Large Binocular Telescope Laboratory. http://oldweb.lbto.org/pdfs/06_Orlando.pdf

- Garner, Robb. (Updated December 27, 2018). James Webb Telescope Overview. Nasa.gov. https://www.nasa.gov/mission_pages/webb/about/index.html

- Crash Course. "Distances #25." YouTube video, 11.20. July 17, 2015. https://www.youtube.com/watch?v=CWMh61yutjU

- Barnett, Amanda, et al. *Orbits and Kepler's Laws.* (Updated June 29, 2020). Nasa Science; Solar System Exploration. Nasa.gov. https://solarsystem.nasa.gov/resources/310/orbits-and-keplers-laws/

- Chamberlin, Alan, and Ryan Park. (Updated February 28, 2014). *Astronomical Unit.* Jet Propulsion Laboratory, California Institute of Technology. Nasa.gov. https://ssd.jpl.nasa.gov/?glossary&term=au

- Doyle, Heather, and Kristen Erickson. (Updated August 27, 2020). *What is a light-year?* Nasa Science: Space Place. Nasa.gov. https://spaceplace.nasa.gov/light-year/en/

- Crash Course. "Distances #25." YouTube video, 11.20. July 17, 2015. https://www.youtube.com/watch?v=CWMh61yutjU

- Ash, Andy. (April 20, 2020). *Why Nasa spacesuits are so expensive.* Insider.com. https://www.businessinsider.com/why-nasa-spacesuits-so-expensive-cost-russia-spacex-2020-3?op=1

- Garcia, Mark. (Updated May 13, 2021). International Space Station Facts and Figures. Nasa.gov. https://www.nasa.gov/feature/facts-and-figures

- Keeter, Bill. (Updated November 5, 2020). *Spot the Station: International Space Station.* Nasa.gov. https://spotthestation.nasa.gov/

- Gupta, Rohit, et al. (June 7, 2021). *Is there gravity in the Space Station?* Brilliant.org. https://brilliant.org/wiki/is-there-gravity-in-the-space-station/

- Engineering Toolbox. (2008). *Hydrogen: Thermophysical Qualities.* The Engineering Toolbox.com. https://www.engineeringtoolbox.com/hydrogen-d_1419.html

- Momsen, Bill. (March 15, 2002. *Mariner IV - First Flyby of Mars.* Wayback Machine, archive. https://web.archive.org/web/20020620141059/http://home.earthlink.net/~nbrass1/mariner/miv.htm

- Mariam-Webster online dictionary. "Cosmonaut." Accessed June 7, 2021. https://www.merriam-webster.com/dictionary/cosmonaut,

- Online Etymology Dictionary. "Cosmonaut." Accessed June 7, 2021. https://www.etymonline.com/word/cosmonaut

- ibid. "Astronaut." https://www.etymonline.com/word/astronaut

- Patel, Neel. (June 14, 2016). *A history of Space Guns.* Inverse.com. https://www.inverse.com/article/16735-a-history-of-space-guns-from-isaac-newton-to-nazis-in-paris-and-project-harp

- Wilson, Jim, editor. (Updated April 13, 2011). *Yuri Gagarin: First Man in Space.* Nasa.gov. https://www.nasa.gov/mission_pages/shuttle/sts1/gagarin_anniversary.html

- May, Sandra. (August 7, 2017). *Who was Alan Shepard?* Nasa.gov. https://www.nasa.gov/audience/forstudents/k-4/stories/nasa-knows/who-was-alan-shepard-k4.html

- Dunbar, Brian. (August 4, 2017). *Profile of John Glenn.* Nasa.gov. https://www.nasa.gov/content/profile-of-john-glenn

- Space.com staff. (March 1, 2012). *How long do footprints last on the Moon?* Space.com. https://www.space.com/14740-footprints-moon.html

- Garcia, Mark. (Updated July 11, 2018). *Sally Ride - First American Woman in Space.* Nasa.gov. https://www.nasa.gov/feature/sally-ride-first-american-woman-in-space

- Mental Floss. "43 Words invented by Authors." YouTube video, 6.53. February 25, 2015. https://www.youtube.com/watch?v=x63y-zV152w, 1:12–1:22

- Brogan, Morris. (May 6, 2021). *10 Great Films about Space Travel.* Bfi.org.uk. .https://www.bfi.org.uk/lists/10-great-films-about-space-travel

- Ormondi, Sharon. (June 4, 2019). *Where does Outer Space begin?* World Atlas.com. https://www.worldatlas.com/articles/where-does-outer-space-begin.html

- Sheth, Khushboo. (June 19, 2019). *Countries with the Most Space Travelers.* World Atlas.com. https://www.worldatlas.com/articles/countries-with-the-most-space-travelers.html

- Brown, Heather. (March 2, 2016). *How many people have gone to Space?* CBS Minnesota. https://minnesota.cbslocal.com/2016/03/02/good-question-astronauts/

- Chepkemoi, Joyce. (June 19, 2019). *What is Space Junk?* World Atlas.com. https://www.worldatlas.com/articles/what-is-space-junk.html

- Drake, Nadia. (December 18, 2020). *Alien Hunters Detect Mysterious Signal from Nearby Star.* National Geographic.com. https://www.nationalgeographic.com/science/article/alien-hunters-detect-mysterious-radio-signal-from-nearby-star

FUN FACTS
SPACE TRIVIA 2.0

PANTHEON SPACE ACADEMY

INTRODUCTION

The universe is enigmatic and rapidly expanding. People have always been fascinated by the mysteries of what lies beyond Earth. Since man first looked up into the night sky, we've been gathering information and learning about the immense space beyond our tiny planet.

How much outer space knowledge do you know? What about your family and friends? Which of you is a true space lover with all the knowledge of the universe? Pantheon Space Academy will test your ability to the limits and inform you about what you don't know. This informative book is perfect for everything from challenging game nights to education and entertainment.

Welcome to Fun Facts Space Trivia Part 2! We will test your knowledge on space exploration, our solar system, and we even have some trivia about the Mars 2020 missions.

But before you join the Space Program, here are 199 trivia questions and 2,001 facts for you to enjoy during a cosmic game night. Challenge yourself or friends and family to see who's memory is out of this world!

CHAPTER ONE

OUR PLANETS AND MOONS

Which moon orbits its planet backward?

Titan, one of Neptune's moons. The unique spin is called a retrograde orbit. Many theories were proposed, but nobody exactly knows why.

Which two planets are the least explored?

Uranus and Neptune. Last visited when Voyager 2 did fly-bys in 1986 and 1989. Although scientists are itching to learn more about Uranus and Neptune, there are no visits scheduled.

Why does Enceladus, one of Saturn's moons, glow?

Enceladus is a small ice moon. Ice and water particles are spewed into its atmosphere and reflect up to 90% of light. It is the most reflective celestial body in our solar system.

Which moon looks like a fictional space weapon and a video game character?

Mimas, one of Saturn's moons. Its appearance is eerily similar to the Death Star, and its heat signature looks like Pac-Man. Luckily it is just a lifeless space rock with no super laser or fear of ghosts.

How many moons does Uranus have?

Uranus has 27 moons. The third most populated after Jupiter and Saturn. Unlike most moons, Uranus' moons are all named after William Shakespeare's characters, such as Ophelia and Puck, and names from Alexander Pope's poems like Belinda and Ariel. Ariel is also a character in Shakespeare's The Tempest.

True/False: The Moon has over 200 tons of human garbage on its surface?

True. The Moon has had many visitors over the years. Dead probes, plants, crashed satellites, backpacks, and even a pair of boots. The Moon has become a landfill, but NASA believes this will give us a unique way of seeing the effects space radiation and the vacuum of space has on these objects.

What are moonquakes?

They are earthquakes on the Moon and can last up to half an hour. They happen because the Sun and Earth play tug-a-war with the Moon. But because the Moon is closer to the Earth, the Earth wins. Some of the quakes were caused by asteroids that hit the Moon. During the Apollo Moon missions, 28 moonquakes were recorded between 1969 and 1977, ranging from 1 to 5 on the Richter scale.

How long would it take to travel to Mars?

Traveling to Mars will take six to eight months, depending on the distance between Earth and Mars at the time of launch.

True/ False: Pluto is smaller than the United States?

True. Pluto is 1,430 miles wide - half the width of the United States.

What is Pluto's Tombaugh Regio?

Pluto's heart-shaped glacier. In 2020, scientists discovered that Pluto's heart is at the center of the planet's atmospheric cycle. It makes the winds blow. And when it snows, the mountains are dressed in red.

How was Uranus discovered?

An accident. Uranus was first observed in the sky in 1690, but astronomers realized it was a planet after William Herschel surveyed nearby stars in 1781. It was also the first planet discovered using a telescope. The element uranium was named after the sphere only eight years later.

What was the name given to Uranus upon its discovery?

Georgium Sidus. Named after King George III of England. For nearly 70 years after its discovery, it was known as planet George.

How long is a day and a year on Uranus?

A Uranian day is 17 hours, and a Uranian year is 84 years. And because of its weird rotation, Uranus winter is 21 years of night, and the summer is 21 years of light.

How was Neptune discovered?

With Math! Johann Gottfried Galle discovered Neptune in 1846. He used mathematical predictions made by John Adams and Jean Le Verrier to determine why Uranus was being pulled out of orbit.

How long is one day on Venus?

It takes Venus 243 days to spin around its axis once, but only 225 Earth days to travel around the Sun. Talk about a year in a day.

If Venus is considered Earth's twin, why are we looking to colonize Mars?

Venus is similar to Earth in both size and composition, true. The big difference is that Venus is our solar system equivalent to hell, and Mars is just a big red desert.

What are the three reasons we can't colonize Venus?

Venus has active volcanoes and lava canals. The surface temperatures are kept at around 869 degrees Fahrenheit by a runaway greenhouse effect and an extra thick layer of carbon dioxide. And, as a bonus, it rains sulphuric acid and molten metal.

Why is Mercury still shrinking?

Its core temperature is slowly cooling down, causing Mercury to shrink and turning its crust into a raisin. Scientists estimate it has contracted around 1.2 to 2.5 miles in the last few millennia.

True/False: Mercury has ice on its surface.

True. NASA's Messenger found ice on Mercury's north pole in craters that have never seen sunlight. Organic materials make a home in these craters, but the planet is too hot and airless for life.

On how many planets can fire burn?

One. Fire can only burn on Earth because no other planet has enough oxygen.

True/False: You can grow crops on Mars, just like in the movie **The Martian.**

True. It is possible to grow vegetables on Mars. But it will be difficult. Just like in the movie, you will need a greenhouse, fertilizer, and, most importantly, water for the plants to grow. But don't expect them to grow as well as in the movie.

How many American flags are on the Moon?

Six Flags. One for each Apollo mission. The following person to walk on the Moon won't see colors on these flags because they have been bleached white by radiation.

What is the most prominent crater on our Moon?

The South Pole-Aitken basin. The crater is 8.1 miles deep and 1,600 miles in diameter.

Which planet is known as the vacuum cleaner of space?

Jupiter. It is so large that it attracts large asteroids that could potentially hit Earth and destroy all life. It acts as our guardian angel to keep Earth safe.

What is Saturn's nickname?

The Jewel of our Solar System. And it's obvious why. Those giant rings are the only ones seen from Earth. Its weather patterns also form color rings on its surface.

How many planets in our solar system are perfectly round?

Only Mercury and Venus. All the other planets look like deflated balls bulging out in the middle.

Which planet, besides our own, has ice caps, and how do they differ from ours?

Mars. Just like Earth. The difference is that Mars' ice caps are seasonal and covered with a layer of carbon dioxide during winter.

What did the Yutu 2 rover find on the far side of the Moon?

In July 2019, it found a strange gel-like green substance. Chinese scientists believe it formed during a volcanic eruption or an impact with another celestial body.

What was Venus known as before its discovery as a planet?

The Morning and Evening Star. Ancient Greeks and Egyptians believed that Venus was two different stars. Venus was known as Phosphoros (the morning star) and Hesperos (the evening star). It was the Hellenistic Greeks who discovered that it was one object.

What compound did scientists recently discover in Venus's atmosphere?

Phosphine. This could be a sign of life in Venus's clouds. Also found on Earth near microbes. The presence of this compound on Venus should be impossible because the atmosphere contains chemicals that destroy phosphine.

What is the next mission planned to visit Venus?

NASA's DAVINCI+ will be the next mission sent to Venus. The aim is to find out how Venus became such a hostile environment and learn more about the long-term effects of the greenhouse effect.

What are objects called that temporarily orbit planets?

A mini-moon. These objects usually hang out in orbit a few months before they are flung back into space.

What 1960s object returned to Earth in 2020?

The rocket booster from a NASA launch. It became one of Earth's mini-moons when it began its orbit around Earth.

True/False: Earth has some of the rarest elements in the universe.

True. Elements such as oxygen and silicon are abundant on Earth and don't exist in space in large quantities. Most other planets and stars are hydrogen and helium.

True/False: Earth is home to Martian rocks which we didn't collect.

True. Asteroids found in the Sahara Desert and Antarctica have been identified as Martian. They have air pockets with air that is a match to the atmosphere of Mars. Where they come from and how long they've been on Earth is still a mystery.

What is the BepiColombo mission?

It's a collaboration between the European Space Agency (ESA) and the Japanese Aerospace Exploration Agency (JAXA) to send two spacecraft to Mercury. It's both agencies' first missions to Mercury and the first to send two orbiters simultaneously.

What was the name of the first spacecraft to orbit an outer planet?

Galileo. It was the first craft to orbit the outer planet, Jupiter, and launch a probe into its atmosphere. It also witnessed the comet Shoemaker-Levy 9 crash into a planet's atmosphere.

How many people worked on the Galileo project from 1989 until it ended in 2003?

Eight hundred people from countries such as the USA, Germany, Britain, France, Canada, and Sweden cost around $112 billion.

CHAPTER TWO

OUR SOLAR SYSTEM

Why do stars twinkle?

Because of the way the light gets disrupted when traveling through Earth's atmosphere, you will hear astronomers call this astronomical scintillation.

When was the Hubble Telescope launched into space?

On April 24, 1990, the Hubble Telescope was launched using the Discovery shuttle. It's as long as a school bus and weighs as much as two elephants. The telescope is solar-powered and sends over 140 gigabytes of raw data to Earth every week!

Are there any interstellar objects in our solar system?

Yes. The first known interstellar traveler was discovered in 2017 and called Oumuamua and most likely came from a planet similar to Pluto. Its name means scout or messenger in Hawaiian. The second traveler is a comet called 2I/Borisov, discovered in 2019.

What is Niku?

Niku is a tiny trans-Neptunian object orbiting the Sun opposite to everything else in the solar system. Why it decided to frustrate scientists is still unknown.

If sound could travel through space, what would the Sun sound similar to?

It would sound like a white-noise rock concert. The Sun is extremely loud, and we would never be able to hear anything else, even at night.

True/ False: The Great Wall of China can be seen from space.

False. The Great Wall can't be seen from space with the naked eye. Yang Liwei, a Chinese astronaut, debunked this myth.

How many lunar eclipses happen a year?

Between two and four lunar eclipses can happen a year, but they are not always visible everywhere, and most are just partial eclipses.

True/False: A comet's tail shows the direction it's traveling in.

False. The tail always faces the Sun, no matter the direction of the comet. The Sun's heat and radiation melt the comet's surface, creating the trail of light.

What is the only known organism that can survive in the vacuum of space?

Tardigrades (water bears or moss piglets) are the only known organisms that can survive in space. They are only less than an inch big and can sustain for up to 10 days unprotected. They are also some of the oldest known life forms on Earth.

What is the hottest layer of the Sun?

The outer layer of the Sun, called the corona, is the farthest layer from the Sun and also the hottest. This is called the inverse temperature effect and is confusing scientists, just like the tiny trans-Neptunian body.

How long does it take the Sun to make a full rotation?

About 25 and 35 days. If you stood on the equator, it would take you 24 days to go in a complete circle. But if you rotated with the Sun at its poles, it would take you up to 35 days.

What is the magnetic bubble around our solar system called?

The Heliosphere. It's created by the Sun and protects us from interstellar radiation.

What is the cycle called during which the Sun reverses polarity?

The solar cycle. The Sun's magnetic field flips, and the Sun's north pole becomes its south pole and vice versa. It happens every 11 years.

What is a massive release of plasma from the Sun called?

A Coronal Mass Ejection. These ejections are the Sun burping, sending shock waves strong enough to interrupt power grids and communications, poison high-flying aircraft with radiation, and be lethal to life in outer space.

Which spacecraft was the first to orbit and land on an asteroid?

NASA's NEAR Shoemaker spacecraft. It was launched in 1996 and entered orbit around the asteroid, Eros. On February 12, 2001, it ended successfully on the asteroid. NASA then lost contact with NEAR on February 28, 2001.

What makes the asteroid Phaeton different from other asteroids?

Phaeton looks and acts like a comet, and it even has a tail. But Phaeton isn't made out of ice and dust like comets. It's a typical asteroid with a tail. When the asteroid moves closer to the Sun, it releases dust that follows it like a tail. These dust particles are responsible for the Geminids meteor shower every December when Phaeton crosses Earth's orbit.

True/False: The asteroid Bennu is on a collision course with Earth and will hit us in 50 years.

False. Bennu will come within 4.6 million miles of Earth in 150 years, and the chances of a collision are 1 in 2,700.

True/False: Sunlight can change Bennu's trajectory.

True. Bennu's days are only 4 hours long, so the side facing the Sun rotates. The Sun warms the asteroid, and when it cools down, it is slowly pushed towards the Sun at around 0.18 miles per year.

The asteroid Bennu is named after which deity?

An ancient Egyptian Sun god linked to creation and rebirth. Bennu is an ancient and primitive asteroid that dates back to the creation of our solar system.

Which NASA spacecraft is on a course to collect an almost four-pound sample from Bennu?

OSIRIS-REx. It was launched in 2016 and reached Bennu in 2018. It's expected back with its massive sample in 2032. Pictures from OSIRIS-REx showed the surface on Bennu was covered in huge rocks, making landing a tricky business.

True/False: Bennu has pieces of the asteroid Vesta, on its surface.

True. Scientists have discovered six large boulders on Bennu that are lighter than the asteroid and seem to be from Vesta. A fragment from Vesta may have struck Bennu's parent asteroid and left behind rocks fused with Bennu when it broke apart.

What is a rubble-pile asteroid, and how are they made?

They are objects made from different pieces of rubble that will fuse together with the help of gravity. It starts when an impact breaks off a much larger entity, and the pieces come together to form a new asteroid. These types of asteroids can effortlessly fall apart again. Bennu is a rubble-pile asteroid the size of the Empire State Building.

True/ False: Asteroids can have rings too.

True. Chariklo, discovered in 1997 by the Spacewatch team at the University of Arizona, is the fifth celestial body to have rings. And the first asteroid.

What is Sedna?

Sedna is a mysterious red planet-like object beyond the Kuiper belt. The closest it comes to the Sun is 3 billion miles from the Kuiper belt. It is possibly the nearest object to us that's part of the Oort Cloud.

What is the line called that marks the start of space?

The Karman Line. It is 62 miles above the Earth. Reaching this line with a car will take less than an hour. Too bad cars can't drive straight up.

What happens when two pieces of metals touch in space?

They spontaneously stick together. This process is called cold-welding and doesn't need an external source for the metals to fuse. This phenomenon was first discovered in the 1940s when scientists did experiments inside a vacuum. During the Gemini IV mission, astronauts were unable to open the door. First thought was that the doors fused closed in outer space. Luckily the door was just stuck.

CHAPTER THREE

BEYOND OUR SOLAR SYSTEM

When will Halley's comet pass Earth again?

2061. Halley's Comet is a periodic comet, meaning it followed a set path. The last time it was near Earth was in 1986. First named after the English astronomer Edmond Halley, Chinese astronomers first observed her in 239 B.C.

What was the first galaxy identified as a spiral?

The Whirlpool Galaxy (Messier 51). Charles Messier discovered it in 1773. Spiral galaxies encompass young hot stars twisted into arms around a supermassive black hole. Most of the known galaxies are spiral.

How many galaxies can you see with the naked eye?

Three. Look below the w shape of Cassiopeia to see Andromeda (M31). M81 and M82, a pair of galaxies, can be found within the Ursa Major constellation.

What is a group of stars that form a picture?

A constellation. They can take the shape of anything but are usually in the frame of a human or mythological creature or an inanimate object like a compass.

Our ancestors used constellations for what purpose?

Navigation. By using the North Star, Polaris, sailors navigated across the oceans.

How many constellations are there in the night sky?

The International Astronomical Union has identified Eighty-eight constellations. Famous constellations include the Zodiac, Ursa Major, Ursa Minor, Orion, and the Heavenly Waters.

How long does it take our solar system to orbit around the Milky Way?

Around 230 million years. It's called a cosmic or galactic year, and the last time the solar system was in its current position, Earth was a giant swamp, and monster amphibians were just beginning to prosper.

What is the biggest cluster of galaxies discovered?

The BOSS Great Wall. The Baryon Oscillation Spectroscopic Survey (BOSS) team discovered this honey-combed structure in 2016, and it contains a supercluster of 830 galaxies.

What is a zombie star?

A zombie star is a star that didn't die. Usually, when a star dies in a supernova, it's gone, but zombie stars survived and went off on another cosmic adventure.

What celestial objects shine so brightly it eclipses its galaxy?

A quasar. These shiny objects seem to live near supermassive black holes at the center of some galaxies. Because they only exist near black holes, studying them can tell us a lot about the universe's history.

What did astronomers find hidden in a quasar?

The biggest and oldest body of water. It contains 140 trillion times more water than can be found on Earth. Because the light coming from this quasar is billions of years old, the water trapped within is only a billion years after the Big Bang.

What is a neutron star?

A neutron star is a super dense object born from a supernova collapsing in on itself. These are so compact, a teaspoon of neutron star particles can weigh almost 900 billion tonnes.

Where can the toughest material in the universe be found?

Inside of a neutron star. It's called nuclear pasta, and it can take over a billion years to make. A team of American and Canadian researchers has determined to break a plate of nuclear pasta; it would take 10 billion times the force needed to shatter steel.

What does the center of the Milky Way smell like?

Raspberries and rum! Sweet, right. Astronomers from the Max Planck Institute used the IRAM radio telescope and found ethyl formate forming in dust clouds. This chemical gives raspberries and rum their flavor. Using chemical analysis, astronomers have found that Titan smells like gasoline, and the Apollo astronauts reported that the Moon smells like gunpowder.

What makes the planet TrES-2b different from the rest of the planets discovered thus far?

It is the darkest planet ever discovered. It's even darker than the darkest paint found on Earth. TrES-2b has no clouds in its atmosphere to reflect sunlight. And it glows red from the extreme temperature of 1,796 Fahrenheit. It's like a glowing piece of coal, only a few light-years away from the constellation Draco, the dragon.

What is the oldest star in the universe?

The Methuselah Star (HD 140283) is around 14 billion years old, found in the Libra constellation. This star is unique because scientists have calculated the universe to be only 13.8 billion years old. Perhaps it knows the secret behind the Big Bang.

What happens when a star dies?

Two things can happen. It will go out in a hail of glory, a supernova, and form a black hole, freezing the dead star in time. Or the star will become a white dwarf and slowly burn out, very anti-climatically.

When was the first photograph of a black hole?

The first photograph of a black hole was made public on 10 April 2019. The black hole is at the center of the Virgo A galaxy and took two years to create from data collected in April 2017 by the Event Horizon Telescope (EHT). They had to send the data to each other via FedEx because the internet can't handle the transfer of such large files. It was a petabyte of data collected!

What is the disk around a black hole called?

An accretion disk. A swirling pool of hot plasma, dust, and gas. It emits X-rays and powerful radiation.

What is a white hole?

A white hole is the opposite of a black hole. Instead of pulling objects in, they spit matter out. Not one white hole has been successfully documented because they spontaneously spew out matter and disappear, unlike a black which stays around for millions of years.

What did the Australian Square Kilometre Array Pathfinder (Askap) reveal when it mapped the universe?

The radio telescope mapped over 3 million new galaxies in just 300 hours. It has 36 dish antennas covering over 86% of the sky and is the first part of an international project to build the world's largest radio telescope, the Square Kilometre Array.

What is a neutron star called with a strong magnetic field?

A magnetar. Its magnetic field is 1000 times stronger than a typical star's magnetic field and will tear anything apart that comes within 600 miles. They're the strongest magnets in the universe, and there are only around 30 that we know of.

When was the first magnetar discovered?

The ultra magnetic star was first observed by two Russian spacecraft in 1979. It wasn't until 2008 that astronomers realized it was a rare type of star.

What are starquakes?

It's a cross between an eruption and a quake that occurs on magnetars. When a magnetar's magnetic field shifts, the crust breaks open, releasing hot plasma from the star's core and sending out a giant burst of radiation energy. Only three of these starquakes have been detected so far in 1979, 1998, and 2004. The starquake in 2004 was so powerful; all satellites went offline for a few minutes and disrupted radio transmissions.

Which exoplanet is mostly water?

Gliese 1214b. The Hubble Space Telescope found water on the exoplanet in 2012 in the form of burning ice.

True/ False: There is a blue exoplanet similar to Earth where it rains glass.

True. It's called HD189733b, first seen by the Hubble telescope in 2013. It looks like Earth, except for one tiny detail. It rains molten glass, sideways, at about 4,300 m/p. The planet's temperature is 1,832 degrees Fahrenheit. It's safe to say this planet doesn't have life.

Which world is appropriately named lava world?

Kepler-78b. It's an Earth-like exoplanet that is 1.2 times the size of Earth. It completes a full orbit of its star once every eight hours and is less than one million miles from its 'sun', making it too hot for any form of life. Unfortunately, the planet is doomed. The star's gravitation will pull it closer and closer until it's burned up.

What is the most distant object we have explored?

Arrokoth, previously known as Ultima Thule. It is located 4 billion miles from Earth and is 4 billion years old. The New Horizons visited it in 2019. Scientists believe Arrokoth formed when two lobes merged in space.

What is the hottest place in the universe?

It is a gas cloud surrounding a cluster of galaxies. The temperature effortlessly reaches over a million degrees Fahrenheit.

What is the most common type of star in the Milky Way?

Red dwarves. They makeup about three-quarters of all the stars in the galaxy and can be seen with the naked eye.

What is the most massive type of star in the universe?

Red supergiants. They are almost a thousand times larger than our Sun and only half as hot. UY Scuti, found in the constellation Scutum, is the closest red giant to Earth; it is so big that if it replaced our Sun, it would engulf Mercury, Venus, Earth, Mars, and Jupiter.

Who said the following quote, "we're all made out of star stuff," and what did he mean?

Carl Sagan. Stars are little factories, creating elements like carbon, oxygen, and hydrogen. These elements are a crucial part of living beings on Earth, and since matter can't be destroyed, we all are made out of little stardust.

True/False: The bigger the star, the shorter its life.

True. The bigger the star is, the more energy it creates, and the faster it burns out. When stars run out of hydrogen for fusion, their core collapses, and they explode into a supernova.

What is the name of the black hole at the center of our galaxy?

Sagittarius A. Classified as a supermassive black hole.

How many layers does the Sun have?

The Sun has three layers and a transition region. The deepest layer visible to us is called the photosphere. The next layer is the chromosphere and is 250 miles above the photosphere and around 14,000 degrees Fahrenheit. Between the chromosphere and the outer layer is the transition region. It's only about 60 miles wide, and the temperature in this region rises sharply from 14,000 degrees to 900,00 degrees Fahrenheit. The Sun's outer layer is called the corona and can only be seen during a total solar eclipse.

If there are more stars than grains of sands, why is our night sky so dark?

Heinrich Olbers, a German astronomer, suggested that starlight is blocked or absorbed before it reaches us, but Edgar Alen Poe gave us a more reasonable answer. He said that our universe is not old enough and too big to be filled with light yet. Starlight from beyond the cosmic horizon hasn't arrived yet.

True/False: You can see the back of your head if you stand in a black hole.

Yes! The light would bounce off the back of your head, go around the black hole, and then back to you—real-life third-person view. Unfortunately, getting so close to a black hole to see yourself in the third person will kill you. Better stick to video games.

Which movie helped scientists understand black holes better?

Interstellar, released in 2014. The visual effects crew wanted smoother pictures for their movie and wrote a code using Einstein's equations. They dubbed this code DNGR and created such an accurate model of the black hole, Gargantua, that scientists now use it to study black holes and their effect on light waves. This code, the DNGR, created such an accurate rendering of the black hole, Gargantua, that scientists now use it to understand black holes better.

What are the fast, invisible ripples called that can squeeze or stretch objects in its path?

Gravitational waves. These are waves of gravity that travel at the speed of light. Documented during supernovas, when two stars orbit each other, or two black holes merge. In 2015 scientists detected a gravitational wave created when two black holes collided and merged 1,3 billion years ago.

True/False: The Universe was much hotter when it was younger.

True. When the universe 'banged' it was over a billion degrees and has steadily cooled down over the millennia. This cooling effect has scientists theorizing that in the end, the universe will completely freeze over.

How much of the universe can we see?

Only 5% of the universe is visible. The rest is invisible dark matter and dark energy.

If dark matter is invisible, how can scientists study it?

Researchers can study them by looking at the effect their gravity has on ordinary matter. Because they weigh six times more than normal matter, they have a greater gravitational pull on the matter around them. They don't interact with magnetic forces, they don't reflect light, but they can bend it. Researchers aren't sure exactly what dark matter is, only that it is dark.

What theory describes how gravity shapes the fabric of space and time?

Einstein's theory of relativity. This theory, published in 1915, led to the discovery of the first black hole in 1964.

True/ False: There is a planet with a crust made out of diamonds.

True. In 2012, scientists discovered an Earth-like planet in the constellation Cancer with a crust made out of diamonds. It's called 55 Cancri e and can be seen with the naked eye. Unfortunately, space mining is still just a dream. Imagine what it would be like to walk around with space bling around your neck.

What is the oldest galaxy in the universe?

BRI 1335-0417 is the oldest known spiral galaxy in the universe. It formed over 12.4 billion years ago, around 2 billion years after the Big Bang. Scientists believe that two galaxies collided to form the galaxy when the universe was still young.

What is the coldest place in the universe?

The Boomerang Nebula. It is -460 degrees Fahrenheit. One degree above absolute zero and three degrees colder than the rest of space. Images from ground-based telescopes showed the nebula as lopsided, but pictures from the Hubble Telescope showed that it's a bow tie. A tad late to rename it now.

What is the most petite star ever discovered?

EBLM J0555-57Ab is about the size of Saturn and is the smallest a star can be. Despite being the same size, they don't have the same temperature. In other words, Saturn is too cold. Can you imagine the solar system without our beautiful ringed planet?

CHAPTER FOUR

AGE OF EXPLORATION

From the Greek words, what is the translation of astronaut?

Star Sailor, from astron, meaning star, and nautes, meaning sailor.

What is the international agreement on peaceful space exploration called?

The Outer Space Treaty. On January the 27th, 1967, a treaty was signed by the United Nations, the Soviet Union, the United States, and several other countries to keep war out of space. The Outer Space Treaty prohibits the use of nuclear or other weapons of mass destruction in outer space. No nation may claim territory in space, and they are responsible for any object launched into space. Also, they are compelled to help astronauts in distress, whatever their nationality.

What does NASA stand for?

National Aeronautics and Space Administration

When was NASA founded?

On July 29, 1958, NASA formed as a civilian agency responsible for America's space research. In response to Russia's Sputnik launch in 1957 and the US-Soviet, the NASA formation would officially begin a "race to the stars."

Who is NASA's oldest female employee?

Susan Finley. She began working at NASA in 1958 and was part of the team of women responsible for writing the code which helped Apollo land on the Moon. Without the help of computers. She wrote some of the code needed to land both Opportunity and Spirit on Mars. She was awarded the NASA Group Achievement award in 2013 and NASA Exceptional Public Service Medal in 2018 for her contributions toward space exploration. When asked when she will retire, she said the following: "I don't plan to stop. I have nothing better to do."

True/False: Snoopy is NASA's official safety Mascot?

True. Snoopy has joined astronauts in space since 1969. NASA also created the Silver Snoopy Award, a silver pin in the shape of Snoopy in his astronaut outfit.

What plants have been grown in space?

Thale cress, purple false brome, Triangle Water Fern, sunflowers, Zinnia, field mustard, mizuna lettuce, zucchini, wheat, red romaine lettuce, rice, tomato, spinach, pepper, broccoli, and radish. A veggie salad with a side of flowers and grass.

What is the Artemis Moon Mission?

It's "humanity's return to the Moon." NASA plans to send the first woman to the Moon by 2024. These missions aim to establish a base on the Moon. Artemis 1 is expected to launch in 2021, Artemis 2 in 2022, and Artemis 3 in 2024. These missions are a giant leap forward for mankind and will set us on course to further explore outer space.

Which probe, launched in 2020, will come closer to the Sun than any probes before?

The Solar Orbiter. This mission is a team effort between NASA and the European Space Agency. It launched in 2020, and its mission is to study the Sun and heliosphere. It will get closer to the Sun than any other probe. Using eleven different state-of-the-art instruments, the Orbiter will be able to create a more detailed picture of how our Sun works. The Orbiter will take two years to reach the Sun, and during that time, it will do two fly-bys past Venus and one past Earth in November 2021. Once in orbit around the Sun, the Solar Orbiter will get as close as 26 million miles to the Sun.

Which shuttle deployed the Hubble Telescope?

The Discovery in 1990. The telescope weighed 24,000 pounds.

What was the first bird to successfully hatch in space?

Japanese Quail. In 1990 Russians successfully hatched five quail eggs aboard the Mir Space Station as part of a study to see the effects of zero-g on developing embryos. They also wanted to determine if the quail would make a viable food source.

What is the ISS?

ISS stands for International Space Station. It's also the most prominent man-made object in space. The space station is the astronauts' home away from home while they are in space. It also contains a laboratory where they can study the effects of zero-g on organic organisms like plants.

When was the ISS launched?

The ISS was launched in pieces and assembled in space. The first piece, the Russian Zarya control module, was launched in November 1998. Astronauts continually added pieces over the next two years before people could start to work from within the ISS.

How big is the ISS?

The ISS is 357 feet in length (almost a football field), weighs over a million tonnes, and has an acre of solar panels powering it.

How long have humans occupied the ISS?

Since November 2nd, 2000. The ISS is the only object besides our planet to have a permanent human presence.

What was the most number of people on board the ISS?

Thirteen. The first time was in 2009, and there have never been more than thirteen people aboard the ISS at a time.

Which man-made objects can be seen from the ISS?

The Great Pyramids of Giza, city lights, the Kennecott Copper Mine, bridges such as the Confederation Bridge and the Mackinac Bridge, Palm Island at Dubai, Almeria's Greenhouses, The Grand Canyon, the Great Barrier Reef, and the Amazon river.

How many space stations are still operational?

Only one, the ISS. The Chinese Tiangong Space Station deorbited in 2019, after a successful mission. The Russian Mir was abandoned when a fungal infestation was growing on the outside windows. It was eating its way through the titanium hull. The spacecraft was destroyed when it reentered our atmosphere in 2001. Good call.

How many manned space stations have orbited Earth?

Ten space stations have been built. They include the US Skylab, the seven Salyut stations, and the Almaz stations.

What is the first object to be designed on Earth and made in space?

A wrench. It was designed on Earth and emailed to a 3D printer onboard the Space Station. It was part of an experiment to test the impact of microgravity on printed tools and to see if it's better to print tools in space rather than to send it up with the next shuttle, which could take months to launch.

True/ False: An astronaut lost his wedding ring on the Moon.

True. Bizarre as it sounds, General Charles Moss Duke, Jr., lost his wedding ring during an 11 day trip on the Moon. He and his fellow crewmates searched for eight days but could not find it. On the ninth day, Charlie found it when he went for a spacewalk. It floated out of the door, hit a fellow astronaut on the back of the head, and floated into Charlie's hand.

Who spent the longest time on the Moon?

Harrison Smith and Eugene Cernan. They spent over three days on the Moon and performed three moonwalks that lasted over 22 hours during Apollo 17's mission.

True/False: Neil Armstrong's iconic quotes were the first words spoken on the Moon.

False. The first words spoken on the Moon were by Buzz Aldrin. His words were "Contact light". Armstrong's quote came 6 hours later when he became the first man to walk on the Moon.

True/False: NASA lost the original footage of the Apollo 11 Moon landing.

True. Rumors started flying around that NASA had lost the tapes and, after an intense search in 2006, they were still lost. NASA concluded that since everything necessary was digitized, these tapes were probably erased and reused. They also stated that there is no missing or unseen footage because the video transmissions were directly converted and broadcasted live on tv.

Who was the first man to be allergic to the Moon?

Jack Schmitt, a geologist who went to the Moon on Apollo 17, was allergic to the Moon. He suffered irritation in his sinuses, nose, eyes, and throat for two hours after taking off his helmet. The culprit: Moondust.

How many people have walked on the Moon during Apollo missions?

Twelve people have been to the Moon on six different Apollo missions. Together they spent over 90 hours on the Moon and did 14 moonwalks.

Why did the Apollo Moon missions stop?

The cost of sending humans to the Moon was astronomical. The first Apollo mission alone cost 20 billion dollars, and the public lost interest in the Apollo missions, which led to a loss of funding.

Who was the last man to walk on the Moon during Apollo missions?

Eugene Cernan was the last man to do a moonwalk. He carved his daughter's name on a Moon rock during the Apollo 17 mission. It will stay there until the Moon ceases to exist, along with footprints and the rover tracks. A meteorite could also destroy it, but the chances of that happening are slim.

True/False: President Nixon wrote a speech if the Moon landing was a failure and the astronauts died.

True. Luckily, Nixon never read the speech. The Moon landing showed us what we're capable of and kickstarted the age of space exploration.

True/ False: A Fisher Space Pen helped Apollo 11 return to Earth.

False. A regular felt-tipped pen, not a space pen, fixed a broken circuit breaker onboard Apollo 11. Astronauts have a plan for everything!

What happened to the collected Moonrocks?

Over 184 Moon rocks from the Apollo 11 and 17 missions have gone missing. Three hundred seventy were collected and given as gifts to other countries. Ireland's Apollo 11 Moonrock, given to the Dunsink Observatory, was found in a landfill after a fire destroyed part of the building in 1974.

What flavor of ice cream was released to commemorate America's Moon landing?

Lunar Cheesecake. It was released in 1969 by Baskin-Robbins and was probably out of this world.

What was the first spacecraft to land on Venus?

The Soviet craft, Venera 7, was the first craft to land successfully on Venus in 1970. It hit the planet at high speed but could still transmit information about the world to Earth.

What caused the near-disaster of Apollo 13?

Fifty-six hours after launch on April 11, 1970, the craft's oxygen tanks onboard Apollo 13 exploded. Despite the impossible rescue mission, all three astronauts made it back to Earth alive.

What caused the Challenger disaster in 1986?

A seal in the right fuel rocket booster failed in the cold air and caused the fuel tank to collapse, and hot gas began to escape. What many believed was an explosion was a giant fireball that tore the spacecraft apart 73 seconds after liftoff.

How many people died on board the Challenger?

Seven. Christa McAuliffe, a teacher, Ronald McNair, Judith Resnik, a NASA astronaut, Ellison Onizuka, Gregory Jarvis, payload specialist, pilot Michael Smith, and commander Dick Scobee. After the tragedy, President Reagan postponed the annual address to the nation, the first and only time it was done, to honor the Challenger's crew. His speech ended with the following quote "We will never forget them, nor the last time we saw them, this morning, as they prepared for their journey and waved goodbye and 'slipped the surly bonds of Earth' to 'touch the face of God.'"

True/ False: Millions of people saw the disaster live on tv?

False. The tragedy occurred in the morning when most people were at work or school. Cable news was relatively new, and the network cut off before the shuttle broke apart. It was mostly children who watched it live. NASA arranged a broadcast to schools because of Christa McAuliffe, who was a teacher.

What was the first shuttle to go into space as part of the Shuttle Space Program?

The Columbia. It weighed 178,000 pounds and was the first of NASA's shuttle capable of flight. It launched on April 12, 1981. It flew 28 missions between 1981 and 2003. It carried the Chandra X-ray Observatory into space and was the shuttle used in 2002 to repair the Hubble Space Telescope. Columbia spent over 300 days in space before it broke apart during re-entry in February 2003 over Texas.

Why did the Columbia space shuttle break apart on re-entry in 2003?

A piece of foam broke off from the external tank and sliced through the spacecraft's wing, and the shuttle disintegrated as cabin pressure was lost. All seven astronauts died in the accident.

Which space shuttle has had the most space missions?

Discovery. It was the third shuttle in space. It went into space 39 times, launched the Hubble Telescope, and helped build the International Space Station. Discovery's first flight was in 1984, and its last was in 2011. It spent over a year in space.

How short was the shortest space flight?

15 minutes. Alan Shepard went into space in NASA's Freedom 7 in 1961 and achieved an altitude of 155 miles.

True/False: When Yuri Gagarin arrived back on Earth, people thought he was an alien.

True. When farmers in rural Russia saw Yuri in his spacesuit, dragging his parachute, they were afraid, and a woman asked Yuri if he was from outer space. He replied, "yes, I am." Before finally telling her that he was Soviet and he needed to find a phone to call Moscow.

How many people have been to space?

There have been 536 people in space who have spent a total of 29,000 days outside of Earth's atmosphere.

Who was the first person to record a song in outer space?

Chris Hadfield, a Canadian astronaut. He did a cover of David Bowies' Space Oddity in 2013.

Who has spent the most time in space?

Gennady Padalka, a Russian cosmonaut, has spent 879 days in space from 1998 to 2015. His first mission was to the Mir space station and commanded the ISS four times.

Who was the first man in space to take a selfie?

Buzz Aldrin. He would become the inspiration for Disney's Buzz Lightyear. He took the selfie during the Gemini 12 mission in 1966.

Who was the oldest man to go into space?

Senator John Hershel Glenn, Jr., was 77 when he went into space aboard the shuttle Discovery in 1998. He was one of seven men chosen to become America's first astronauts in 1959.

Who did the most spacewalks?

Anatoly Solovyev, a Russian cosmonaut. He spent over 82 hours in space outside his spacecraft.

Who sent the first tweet from space?

Michael J. Massimino sent a tweet from space on the 11th of May 2011 from the shuttle Atlantis. Since then, astronauts from NASA have been using Twitter to share their lives on the ISS.

What is the furthest distance humans have been from Earth?

248,655 miles. This record was set during Apollo 13 to the Moon in 1970.

How do astronauts eat in space?

Fifty years ago, they had to suck food paste through straws. Now they can eat almost anything in space. All food aboard spacecraft is freeze-dried, and beverages are dehydrated, so astronauts only have to add water. They eat three times a day and have snacks. It takes about 20 to 30 minutes to heat and hydrate a meal.

How do astronauts keep items secure in space?

They use Velcro. Velcro, invented by George de Mestral, is widely used by astronauts in space. They even have Velcro patches in their helmets to help with itches.

What was the first commercial soft drink in space?

Coca-Cola and Pepsi. They spent millions of dollars to create a way for astronauts to drink the beverage without making a mess. When Pepsi got wind of this, they joined the race to invent a way for soda consumption in space. This became known as the 'space cola wars'.

Why do astronauts have to sleep next to fans in outer space?

To keep from suffocating. In zero-g, CO_2 exhaled by astronauts forms a cloud in front of their face, and without ventilation, they will asphyxiate on their exhaled air.

How long can an astronaut survive in space without a spacesuit?

Humans can survive for minutes in the vacuum of space without a spacesuit but will be unconscious in less than a minute.

True/False: Russia's space pen is a pencil.

False. Pencils can burn, and a broken tip is a potential electrical hazard. NASA developed the space pen with a budget of less than $1 million that could write in the vacuum of space.

How much did the ACES spacesuit weigh?

It weighed 280 pounds and took 45 minutes to put on. It wasn't meant for emergencies. With some help from SpaceX, NASA has designed a new, sleeker white spacesuit to replace the old bulky pumpkin suits.

What is the new spacesuit called?

Starman suits. These are one-piece white suits with 3-D printed helmets and touch-screen sensitive gloves. Jose Fernandez, a movie costume designer, said the new spacesuits look straight from a movie set. SpaceX's crew used these during the Dragon test flight. But as futuristic as they may sound, they are for spacewalks. The aliens will just have to come aboard to see the newest in space fashion.

Which children's toy went into space?

Disney's Space Ranger: Buzz Lightyear. A Buzz Lightyear action figure joined the crew in 2008 onboard the ISS and returned 15 months later. The action figure is part of an exhibition at the Air and Space Museum.

How fast does a shuttle travel in orbit?

17,500 mph. Crew members onboard experience a sunset or sunrise every 45 minutes.

What is the Kessler Syndrome?

The Kessler Syndrome may stop space travel and keep us on Earth. It is a theoretical scenario where space junk forms a barrier around Earth and prevents us from leaving Earth. This may become a massive problem because we keep sending stuff up and not bringing it back, but perhaps keep us safe from any hostile aliens.

What is the new project called that will help clean up space debris?

It's called the End-of-Life Services by Astroscale demonstration. It's part of an experiment to clean up all the space debris orbiting the planet, risking the future of space exploration and astrophotography.

Why are launch pads soaked just before a shuttle launch?

To protect us from sound damage. The water acts as a sound suppressor and prevents fire from the extreme temperature of the engines. The 'smoke' from the engines is steam.

How many tiles are on the outside of space shuttles?

21,00 lightweight tiles. These custom tiles are designed to be lightweight and heat reflecting.

True/ False: A microwave oven baffled scientists using the Radio Telescope in Australia.

True. Scientists have had trouble with interference since 1998 and believed lightning to be the cause. But in 2015, they discovered it was the microwave in the staff room at the Parkes Observatory being opened before it finished heating.

What is the difference between a rover and a space probe?

A space probe travels through space and collects data to send back to Earth. A rover is a vehicle designed for ground exploration on planets.

Who are Tom and Jerry?

No, not the cartoon characters. Tom and Jerry are the nicknames of two satellites 'chasing' each other as part of a NASA mission to understand Earth's climate.

How many miles did the Messenger spacecraft travel?

About 4.9 billion miles. It launched in 2004, orbiting the Sun 15 times at an average speed of 84,500 mph. It is the first spacecraft to orbit Mercury since Mariner 10 and is about the size of an office desk with solar pane wings. It has seven scientific instruments to map the planet's surface and study its composition and magnetic field.

Why hasn't the Sun's heat melted Messenger?

It has a heat-resistant and reflective parasol that can withstand up to 700 degrees F while keeping the craft at a low temperature of 70 degrees F.

True/False: Mars is the only planet inhabited by robots.

True. Mars is the only planet that is home to only robots. The world is currently home to six from NASA - Sojourner, Spirit, Opportunity, Curiosity, Perseverance and Ingenuity, and Zhurong from China.

How long was the rover Opportunity supposed to last on Mars?

Only 90 days, but the little rover lasted for 15 years before finally being lost in a dust storm on February 13, 2019. Opportunity taught us that Mars has a wet past and revealed uncharted terrain. Scientists sent over 835 recovery commands to try and get it back online, but the little rover never responded.

True/False: The Mars rover, Curiosity plays Happy Birthday to itself.

True, the lonely rover has wished itself a happy birthday since August 5, 2012. Can we give it a hug?

How long has the rover Curiosity been on Mars?

Curiosity has been on Mars for over nine years. It landed on the Red Planet on August 5, 2012.

What happened to the Mars rover, Spirit?

Six years after Spirit landed on Mars, one of her wheels got stuck in a sand patch in 2010. NASA tried to help her escape, but Spirit stopped responding after a few weeks of becoming trapped. Before Spirit could even leave her landing shell behind, she discovered the first traces of Mars's wet past.

When did Sojourner land on Mars?

Sojourner was the first rover on Mars. It landed in 1997 and took over 550 photos of the Red planet. After Pathfinder's transmitter died, Sojourner cannot communicate with Earth and is still on Mars' surface.

How far had the Mars rover, Opportunity, traveled before she lost power?

Opportunity traveled more than 28 miles on Mars's surface, during which she studied the craters Victoria and Endeavor. She was caught in a dust storm in the Perseverance Valley in 2018 and could not charge her solar batteries. NASA declared Opportunity dead in February 2019.

What did Spirit and Opportunity have in common, apart from being rovers on Mars?

They both had pieces of the World Trade Center fashioned into shields to protect their drills.

When was Perseverance launched, and when did it land and Mars?

NASA's newest rover was launched on July 30, 2020, from Florida and landed on Mars on February 18, 2021.

What is Perseverance's primary mission?

Perseverance's main task is searching for signs of microbial life on Mars. It is outfitted with instruments to collect and store samples which will be retrieved later. Perseverance is also testing technologies that will help humans live on Mars.

Why did Perseverance land in the Jezero Crater on Mars?

NASA chose the Jezero Crater as the best place to start looking for signs of ancient life on Mars since the evidence shows that the crater was once home to an ancient lake. If microbial life once existed on Mars, the best place to find them would be in the sediment remains in the crater.

How is NASA reaching out to any life that might exist on Mars?

NASA gave each rover an object as a memorial and symbol of life on Earth. The tradition is known as festooning. The practice started with the first space probe launch. The Curiosity rover carried a 1909 penny, for geologists, to reference scale. Opportunity and Spirit took parts of the World Trade Center and brought a memorial to the crew of the Columbia. Perseverance packed a sundial with images of bacteria, plants, dinosaurs, and humans and has the following sentence engraved behind one of her cameras "Are we alone? We came here to look for signs of life and collect samples of Mars for study on Earth. To those who follow, we wish a safe journey and the joy of discovery."

Which spacecraft gave us a global view of Mars's discrete auroras?

The Hope spacecraft, belonging to the United Arab Emirates, took the first global pictures of Mars's night lights in June 2021. The spacecraft orbits Mars at a high altitude to study the planet from a new point of view. The discrete auroras are similar to our Northern lights, except they cover the entire sphere.

How many countries are currently studying Mars with probes and rovers?

Three. The US, China, and the United Arab Emirates. For China and the UAE, it's their first mission to another planet. China's Tianwen-1 mission has sent an orbiter, a lander, and a rover. The UAE has sent their Hope Orbiter.

Who was the first tourist in space?

Dennis Tito. He was 60 years old when he went into space. He stayed a week onboard the ISS in 2001.

When is the first space hotel scheduled to open?

The Voyager Station, previously known as the Von Braun Station, is scheduled to open in 2027. Construction will begin in 2026. The designers' aim at Orbital Assembly Corporation is to have guests enjoy space in comfort with regular beds and show lounge bars.

Who is Wally Funk?

The oldest person to fly in space! Wally Funk is an 82-year-old woman who finally got her shot. She was one of the first women trained in the 1960s to go to space but never went. She was hand-picked by Jeff Bezos. Funk joined the multi-millionaire in the New Shepard reusable rocket for a 10 minute trip on July 20, 2021, on the 52nd anniversary of the Apollo 11 Moon landing.

Who will be the first actor in space?

It looks like the Russians will win this race. The actress Yulia Peresild and Klim Shipenko, director and actor, will be leaving for the ISS on the 5th of October 2021. Tom Cruise also intends to depart for space in October 2021, backed by NASA and SpaceX's Elon Musk. The race to be the first star amongst the stars is jokingly called The Space Race 2.

What does a Planetary Protection Officer do?

Their most important mission is that microbes from Earth don't contaminate other planets or moons. Scientists ensure that any probe or rover that's going to land on another planet is sterile. Their second objective is to ensure Earth isn't contaminated by rocks and dust from outer space. Thanks to them, we know any samples from outer space with microbial life didn't come from us.

What is the U.S Space Force?

They are the 6th branch of the US military trained to protect US interests in space, mainly the military satellites, test new space technologies, and develop the Theory of Space Warfighting. Sounds cool, right? But the USSF doesn't carry plasma blasters and stay in Constitution Class battleships to fight off an alien invasion, yet.

FINAL WORDS

Thank you for joining us on another fact-filled adventure through space. Being a scientist isn't about sitting in a lab with test tubes. It's about going out there and discovering. You don't have to go to Outer Space for that; you can start in your backyard.

"We make our world significant by the courage of our questions and the depth of our answers" - Carl Sagan.

REFERENCES

Chapter One: Planets and Moons

Bartels, M. (2019, May 29). *Scientists Want to Probe Atmospheres of Uranus and Neptune.* Space.com. https://www.space.com/uranus-neptune-atmosphere-probe-studies.html

Castro, J. (2015, February 3). *What Would It Be Like to Live on Venus?* Space.com. https://www.space.com/28357-how-to-live-on-venus.html

Cessna, A. (2009, August 10). *Mythology of the Planets.* Universe Today. https://www.universetoday.com/37122/mythology-of-the-planets/

Charles Q. Choi. (2017, November 14). *Dwarf Planet Pluto: Facts About the Icy Former Planet.* Space.com. https://www.space.com/43-pluto-the-ninth-planet-that-was-a-dwarf.html

Cool Cosmos. (2020). *When was Neptune discovered?* Cool Cosmos. https://coolcosmos.ipac.caltech.edu/ask/146--When-was-Neptune-discovered-

Earthsky. (2019, March 13). *Today in science: Uranus discovered by accident | Space | EarthSky.* Earthsky.org. https://earthsky.org/space/this-date-in-science-uranus-discovered-completely-by-accident/

Earthsky. (2020a, February 9). *EarthSky | Pluto's icy heart makes winds blow.* Earthsky.org. https://earthsky.org/space/plutos-icy-heart-makes-winds-blow/

Geggel, L. (2018, May 2). *How Much Trash Is on the Moon?* Livescience.com. https://www.livescience.com/61911-trash-on-moon.html

Howell, E. (2018, July 24). 8 *Cool Destinations That Future Mars Tourists Could Explore.* Space.com; Space. https://www.space.com/41254-touring-mars-red-planet-road-trip.html

Knight, J. D. (2016). *Triton, moon of Neptune - The Solar System on Sea and Sky.* Seasky.org. http://www.seasky.org/solar-system/neptune-triton.html

Mission Juno. (n.d.). *Jupiter's Influence.* Mission Juno. https://www.missionjuno.swri.edu/origin?show=hs_origin_story_jupiters-influence

NASA. (2018, September 25). *Enceladus | Science – Solar System Exploration: NASA Science.* Solar System Exploration: NASA Science. https://solarsystem.nasa.gov/missions/cassini/science/enceladus/

NASA. (2019c, July 28). *In Depth | BepiColombo – NASA Solar System Exploration.* NASA Solar System Exploration. https://solarsystem.nasa.gov/missions/bepicolombo/in-depth/

NASA. (2019d, February 21). *In Depth | Moons – NASA Solar System Exploration.* NASA Solar System Exploration. https://solarsystem.nasa.gov/moons/in-depth/

NASA. (2019b, February 14). *Galileo - Overview.* NASA Solar System Exploration. https://solarsystem.nasa.gov/missions/galileo/overview/

NASA. (2005). *Saturn Fun Facts.* Nasa.gov. https://www.nasa.gov/audience/forstudents/k-4/home/F_Saturn_Fun_Facts_K-4.html

NASA. (2019a, January 28). *Uranus.* Solar System Exploration: NASA Science. https://solarsystem.nasa.gov/planets/uranus/overview/

NASA. (2019c). *Venus 3D Model.* Nasa.gov. https://spaceplace.nasa.gov/all-about-venus/en/

Space.com. (2012, November). *Saturn's Second "Pac-Man" Moon Revealed in NASA Photos.* Space.com. https://www.space.com/18634-saturn-pac-man-moons-photos.html

RMG. (n.d.). *Interesting facts about the Moon.* Royal Museums Greenwich. https://www.rmg.co.uk/stories/topics/interesting-facts-about-moon

Wall, M. (2011, March 16). *10 Surprising Facts About NASA's Mercury Probe.* Space.com. https://www.space.com/11147-nasa-mercury-spacecraft-surprising-facts-messenger.html

Weitering, H. (2016, December 15). *Actually, That IS a Moon: Saturn's "Death Star"- Like Mimas.* Space.com. https://www.space.com/35036-saturn-death-star-moon-mimas-explained.html

Chapter Two: Our Solar System

Chung, E. (2015, July 28). *Beyond Pluto: 5 things left to explore in our solar system.* CBC. https://www.cbc.ca/news/science/beyond-pluto-5-things-left-to-explore-in-our-solar-system-1.3167012

Earthsky. (2018, March 26). *10 surprises about our solar system | Space | EarthSky.* Earthsky.org. https://earthsky.org/space/ten-things-you-may-not-know-about-the-solar-system/

Grossman, D. (2016, August 11). *Meet Niku, the Weird Object Beyond Neptune That Nobody Can Figure Out.* Popular Mechanics. https://www.popularmechanics.com/space/deep-space/a22293/niku-weird-object-beyond-neptune/

NASA. (2019e, December 19). *In Depth | 10199 Chariklo.* NASA Solar System Exploration. https://solarsystem.nasa.gov/asteroids-comets-and-meteors/asteroids/10199-chariklo/in-depth/

NASA. (2019d, July 31). *In Depth | NEAR Shoemaker – NASA Solar System Exploration.* NASA Solar System Exploration. https://solarsystem.nasa.gov/missions/near-shoemaker/in-depth/

Michelsohn, N. (2020, April 28). *The Asteroid That Acts Like a Comet.* NASA. https://www.nasa.gov/feature/the-asteroid-that-acts-like-a-comet

Pavlovic, N. (2016, April 10). *150 Cool Space Facts 99% of People Will Never Know.* The Daily Research. https://www.thedailyresearch.com/space-facts/4/

Scharping, N. (2020, February 4). *What would the Sun sound like?* Astronomy.com. https://astronomy.com/news/2020/02/what-would-the-sun-sound-like

Steigerwald, B. (2020a, July 31). *OSIRIS-REx Finds Vesta Meteorites on Asteroid Bennu.* NASA. https://www.nasa.gov/feature/goddard/2020/bennu-vesta-meteorites

TKSST. (2016, October 28). *Welding in Space.* The Kid Should See This. https://thekidshouldseethis.com/post/welding-in-space-veritasium-gemini-4

Whitt, K. K. (2021, February 24). *Best observing targets for binoculars | Astronomy Essentials* Earthsky.org. https://earthsky.org/astronomy-essentials/best-targets-for-binoculars-moon-planets-nebula-clusters/

Young, C. A. (2014, February 3). *The Sun's Magnetic Poles Have Flipped.* The Sun Today with C. Alex Young, Ph.D. https://www.thesuntoday.org/solar-facts/suns-magnetic-poles-flipped-solar-max-is-here/

Zell, H. (2015, March 2). *Solar Rotation Varies by Latitude*. NASA. https://www.nasa.gov/mission_pages/sunearth/science/solar-rotation.html

Chapter Three: Beyond Our Solar System

AMNH. (2020). *Olber's Paradox: Why Is the Sky Dark at Night?* American Museum of Natural History. https://www.amnh.org/exhibitions/journey-to-the-stars/educator-resources/stars/olbers-paradox

Baker, H. (2021, May 24). *Oldest spiral galaxy in the universe captured in fuzzy photo*. Livescience.com. https://www.livescience.com/oldest-spiral-galaxy-in-universe.html

BBC News. (2020b, December 1). *Australian scientists map millions of galaxies with new telescope*. BBC News. https://www.bbc.com/news/world-australia-55139976

BBC. (2020, October 9). *7 galaxies to observe in the night sky*. BBC Sky at Night Magazine. https://www.skyatnightmagazine.com/advice/skills/best-galaxies-observe-night-sky/

Choi, C. Q. (2011, August 11). *Coal-Black Alien Planet Is Darkest Ever Seen*. Space.com. https://www.space.com/12612-alien-planet-darkest-coal-black-kepler.html

Choi, C. Q. (2021, March 22). *Interstellar object 'Oumuamua is a pancake-shaped chunk of a Pluto-like planet.* Space.com. https://www.space.com/interstellar-object-oumuamua-pancake-shape-pluto-like-planet

Constellation Guide. (n.d.). *Constellation Families.* Www.constellation-Guide.com. https://www.constellation-guide.com/constellation-names/constellation-families/

Earthsky. (2011a, May 27). *Do white holes exist?* Earthsky.org. https://earthsky.org/space/have-we-seen-a-white-hole/

Earthsky. (2011b, July 25). *The biggest oldest body of water in universe.* Earthsky.org. https://earthsky.org/space/largest-oldest-mass-of-water-in-universe-discovered/

Earthsky. (2016, November 28). *How long to orbit Milky Way's center?* Earthsky.org. https://earthsky.org/astronomy-essentials/milky-way-rotation/

Earthsky. (2020b, February 12). *What is a neutron star?* Earthsky.org. https://earthsky.org/astronomy-essentials/definition-what-is-a-neutron-star/

Earthsky. (2021, June 13). *What is a magnetar?* Earthsky.org. https://earthsky.org/space/what-is-a-magnetar/

Frank, J. (n.d.). *Accretion disk | astronomy.* Encyclopedia Britannica. https://www.britannica.com/science/accretion-disk

Garner, R. (2017, October 6). *Messier 51 (The Whirlpool Galaxy)*. NASA. https://www.nasa.gov/feature/goddard/2017/messier-51-the-whirlpool-galaxy

Howell, E. (2015, December 23). *How Big Is The Milky Way?* Universe Today. https://www.universetoday.com/75691/how-big-is-the-milky-way/

Howell, E. (2017a, September 20). *Halley's Comet: Facts About the Most Famous Comet*. Space.com. https://www.space.com/19878-halleys-comet.html

Nagaraja, M. P. (2009). *The Big Bang*. Nasa.gov. https://science.nasa.gov/astrophysics/focus-areas/what-powered-the-big-bang

NASA. (2008). *Black Holes | Science Mission Directorate*. Nasa.gov; NASA. https://science.nasa.gov/astrophysics/focus-areas/black-holes

NASA. (2019a). *Black Hole Image Makes History; NASA Telescopes Coordinate Observation*. NASA. https://www.nasa.gov/mission_pages/chandra/news/black-hole-image-makes-history

NASA. (2015). *What Is a Gravitational Wave? | NASA Space Place – NASA Science for Kids*. Nasa.gov. https://spaceplace.nasa.gov/gravitational-waves/en/

Nield, D. (2019, July 19). *Astronomers Found 3 "Zombie" Stars That Came Back to Life After Supernova.* ScienceAlert. https://www.sciencealert.com/astronomers-have-found-three-zombie-stars-that-came-back-to-life-after-supernova

Redd, N. T. (2017a, November 7). *Einstein's Theory of General Relativity.* Space.com. https://www.space.com/17661-theory-general-relativity.html

Redd, N. T. (2017b, November 14). *What is Dark Matter?* Space.com. https://www.space.com/20930-dark-matter.html

Redd, N. T. (2018a, February 24). *Quasars: Brightest Objects in the Universe.* Space.com; Space. https://www.space.com/17262-quasar-definition.html

Redd, N. T. (2018b, July 26). *What Is the Biggest Star?* Space.com. https://www.space.com/41290-biggest-star.html

Redd, N. T. (2018c, November 19). *What Is a Spiral Galaxy?* Space.com; Space. https://www.space.com/22382-spiral-galaxy.html

Rojo, S. (2018, September 20). *Interesting fact of the month.* NASA. https://www.nasa.gov/ames/spacescience-and-astrobiology/interesting-fact-of-the-month

Sabin, D. (2016, January 11). *Starquake!* Scienceline. https://scienceline.org/2016/01/starquake/

Sokol, J. (2016, March 8). *Billion-light-year galactic wall may be largest object in cosmos.* New Scientist. https://www.newscientist.com/article/2079986-billion-light-year-galactic-wall-may-be-largest-object-in-cosmos/

Specktor, B. (2018, September 19). *The Strongest Material in the Universe Could Be (Nuclear) Pasta.* Livescience.com. https://www.livescience.com/63619-nuclear-pasta-strongest-substance.html

Stickler, J. (2020, January 9). *Where is the coldest place in the universe?* ZME Science. https://www.zmescience.com/science/coldest-place-universe-90534/

Team. (2015, December 3). *The Milky Way smells of rum and tastes like raspberries.* How It Works. https://www.howitworksdaily.com/the-milky-way-smells-of-rum-and-tastes-like-raspberries/

Villanueva, J. C. (2015, December 25). Interesting Facts About the Universe - Universe Today. Universe Today. https://www.universetoday.com/37927/interesting-facts-about-the-universe/

Wall, M. (2015, February 13). *"Interstellar" Visual Effects Team Publishes Black Hole Study.* Space.com. https://www.space.com/28552-interstellar-movie-black-holes-study.html

Wall, M. (2013, October 30). *Strange "Lava World" Is Most Earthlike Alien Planet Yet.* Space.com. https://www.space.com/23394-strange-alien-planet-earthlike-kepler-78b.html

Wall, M. (2019, November 12). *Meet Arrokoth: Ultima Thule, the Most Distant Object Ever Explored, Has a New Name.* Space.com. https://www.space.com/ultima-thule-beyond-pluto-new-name-arrokoth.html

Chapter Four: Age of Exploration

Ahmed, I. (2019, July 16). *At 82, NASA pioneer Sue Finley still reaching for the stars.* Phys.org. https://phys.org/news/2019-07-nasa-sue-finley-stars.html

Aliana. (2019, May 16). *Snoopy, Charlie Brown and the Apollo 10 mission.* Www.kennedyspacecenter.com.https://www.kennedyspacecenter.com/blog/snoopy-charlie-brown-and-apollo-10

Atkinson, N. (2012, February 7). *Can you See the Pyramids from Space?* Universe Today. https://www.universetoday.com/93398/can-you-see-the-pyramids-from-space/

Barbier, R., & 2020. (2020, July 23). *The Purpose and Mission of the Space Force.* American University. https://www.american.edu/sis/centers/security-technology/the-purpose-and-mission-of-the-space-force.cfm

BBC News. (2020a, May 28). *Nasa SpaceX launch: Evolution of the spacesuit.* BBC News. https://www.bbc.com/news/science-environment-52787365

BBC News. (2015, September 12). *Russian cosmonaut record-breaker Padalka returns to Earth.* BBC News. https://www.bbc.com/news/science-environment-34231700

BBC News. (2021, March 22). *Space debris removal demonstration launches.* BBC News. https://www.bbc.com/news/science-environment-56482726

Bosworth, M. (2012, February 20). *What has happened to Nasa's missing Moon rocks?* BBC News. https://www.bbc.com/news/magazine-16909592

Britannica. (2018). *Outer Space Treaty | 1967.* Encyclopædia Britannica. https://www.britannica.com/event/Outer-Space-Treaty

Chang, K. (2021, June 30). *Mars Has Auroras and a U.A.E. Spacecraft Captured New Pictures of Them.* The New York Times. https://www.nytimes.com/2021/06/30/science/mars-aurora-uae.html

Collect Space. (2020, August 11). *"Space cola wars" When Coca-Cola, Pepsi tested soda in space* | collectSPACE. CollectSPACE.com. http://www.collectspace.com/news/news-081120a-space-cola-wars-35-years.html

Crookes, D. (2019, October 16). *How Can a Star Be Older Than the Universe?* Space.com; Space. https://www.space.com/how-can-a-star-be-older-than-the-universe.html

Dakwala, R. (2018, October 18). *NASA's GRACE-FO mission creates new possibilities for climate change research.* The Daily Texan. https://thedailytexan.com/2018/10/21/nasas-grace-fo-mission-creates-new-possibilities-for-climate-change-research/

Dick, S., J. (2011). *The Birth of NASA.* Nasa.gov. https://www.nasa.gov/exploration/whyweexplore/Why_We_29.html

Harbaugh, J. (2015, March 18). *Space Station 3-D Printer Builds Ratchet Wrench To Complete First Phase.* NASA. https://www.nasa.gov/mission_pages/station/research/news/3Dratchet_wrench

History.com. (2020, October 27). *John Glenn returns to space.* HISTORY. https://www.history.com/this-day-in-history/john-glenn-returns-to-space

How Stuff Works. (2008, March 10). *How do astronauts eat in space?* HowStuffWorks. https://science.howstuffworks.com/astronauts-eat-in-space.htm

Howell, E. (2017b, November 14). *Columbia Disaster: What Happened, What NASA Learned.* Space.com. https://www.space.com/19436-columbia-disaster.html

Howell, E. (2017c, December 11). *Discovery: NASA's Busiest Shuttle.* Space.com. https://www.space.com/18187-space-shuttle-discovery.html

Howell, E. (2020, September 18). *Here's every successful Venus mission humanity has ever launched.* Space.com. https://www.space.com/venus-mission-success-history

Howell, E. (2017, August 19). *Here's What It's Like to Be the Planetary Protection Officer at NASA.* Space.com. https://www.space.com/37862-heres-what-its-like-to-be-the-planetary-protection-officer-at-nasa.html

Howell, E. (2020a, May 22). *How SpaceX's sleek spacesuit changes astronaut fashion from the space shuttle era.* Space.com. https://www.space.com/spacex-crew-dragon-spacesuits-explained.html

Howell, E. (2020b, December 31). *These are the space missions to watch in 2021.* Space.com. https://www.space.com/space-missions-to-watch-in-2021

Jenner, L. (2021, June 2). *NASA to Explore Fate of Earth's Mysterious Twin with Goddard DAVINCI+*. NASA. https://www.nasa.gov/feature/goddard/2021/nasa-to-explore-divergent-fate-of-earth-s-mysterious-twin-with-goddard-s-davinci

Jordan, G. (2011). *Can Plants Grow with Mars Soil?* NASA. https://www.nasa.gov/feature/can-plants-grow-with-mars-soil/

KGO. (2016, November 14). *Buzz Aldrin shares first "selfie" taken in space*. ABC7 San Francisco. https://abc7news.com/buzz-aldrin-nasa-space-neil-armstrong/1606315/

Letzer, R. (2021, January 2). *9 epic space discoveries you may have missed in 2020*. Livescience.com. https://www.livescience.com/epic-space-discoveries-of-2020.html

Mars One. (2019). *How long does it take to travel to Mars? - A Mission to Mars*. Mars One. https://www.mars-one.com/faq/mission-to-mars/how-long-does-it-take-to-travel-to-mars

McMahon, C. (2018, August 16). *Can You See the Back of Your Own Head?* Scientific Scribbles. https://blogs.unimelb.edu.au/sciencecommunication/2018/08/16/can-you-see-the-back-of-your-own-head/

Mendelson, Z. (2016, October 6). *That Time an Astronaut Lost His Wedding Ring in Space*. Wired. https://www.wired.com/2016/06/time-astronaut-lost-wedding-ring-space/

NASA. (2020, December 8). *5 Hidden Gems Are Riding Aboard NASA's Perseverance Rover*. NASA's Mars Exploration Program. https://mars.nasa.gov/news/8812/5-hidden-gems-are-riding-aboard-nasas-perseverance-rover/

NASA. (n.d.). *Landing Site*. Mars.nasa.gov. https://mars.nasa.gov/mars2020/mission/science/landing-site/

NASA. (2017). *Overview - Mars 2020 Rover*. Nasa.gov. https://mars.nasa.gov/mars2020/mission/overview/

NASA. (2019b). *Mars 2020 Rover*. Nasa.gov. https://mars.nasa.gov/mars2020/

NASA. (2020, October 30). *What Is the International Space Station?* NASA. https://www.nasa.gov/audience/forstudents/5-8/features/nasa-knows/what-is-the-iss-58.html

OSS. (2017, May 24). *Sleeping Astronauts*. Office for Science and Society. https://www.mcgill.ca/oss/article/did-you-know-general-science/sleeping-astronauts

Pruitt, S. (n.d.). *What Went Wrong on Apollo 13?* HISTORY. https://www.history.com/news/apollo-13-what-went-wrong

Pruitt, S. (2018, October 19). *5 Things You May Not Know About the Challenger Shuttle Disaster.* HISTORY. https://www.history.com/news/5-things-you-might-not-know-about-the-challenger-shuttle-disaster

Rabie, P. (2019, July 11). NASA *Addresses Controversy Over "Lost Tapes" of Apollo 11 Moonwalk.* Space.com. https://www.space.com/nasa-apollo-11-moonwalk-lost-tapes-auction-statement.html

Rehm, J. (2018, October 10). *What Is the U.S. Space Force?* Space.com. https://www.space.com/42089-space-force.html

Rhawn, G. J. (2017, February 24). *Cosmology.com.* Cosmology.com. http://cosmology.com/SpaceFungi.html

RMG. (n.d.). *Why did we stop going to the Moon?* Www.rmg.co.uk. https://www.rmg.co.uk/stories/topics/why-did-we-stop-going-moon

Shen, A. (2012, March 29). Buzz Lightyear: To Infinity, and the Air and Space Museum. Smithsonian Magazine. https://www.smithsonianmag.com/smithsonian-institution/buzz-lightyear-to-infinity-and-the-air-and-space-museum-167687292/

Shoard, C. (2021, May 13). *Space race 2: Russian actor bound for ISS in same month as Tom Cruise.* The Guardian. https://www.theguardian.com/science/2021/may/13/russia-send-actor-director-iss-shoot-first-movie-space

Skinner, J. (2008, October 14). *Life on Mars.* Torro Magazine.

Sky News. (2021, July 1). *Wally Funk: Woman, 82, gets chance to go into space 60 years after missing out due to her gender.* Sky News. https://news.sky.com/story/wally-funk-woman-82-gets-chance-to-go-into-space-60-years-after-missing-out-due-to-her-gender-12346764

Stillman, D. (2015). *What Is the Hubble Space Telescope?* NASA. https://www.nasa.gov/audience/forstudents/5-8/features/nasa-knows/what-is-the-hubble-space-telecope-58.html

Street, F. (2021, April 30). *First space tourist: "It was the greatest moment of my life."* CNN. https://edition.cnn.com/travel/article/space-tourism-20-year-anniversary-scn/index.html

Street, F. (2021, March 5). *World's first space hotel scheduled to open in 2027.* CNN. https://edition.cnn.com/travel/article/voyager-station-space-hotel-scn/index.html

Strickland, A. (2019, February 14). *After 15 years, the Mars Opportunity rover's mission has ended.* CNN. https://edition.cnn.com/2019/02/13/world/nasa-mars-opportunity-rover-trnd/index.html

Tan, M. (2015, May 5). *Microwave oven to blame for mystery signal that left astronomers stumped.* The Guardian. https://www.theguardian.com/science/2015/may/05/micr owave-oven-caused-mystery-signal-plaguing-radio-telescope-for-17-years

Thompson, A. (2021, January 21). *The Kessler Syndrome.* The National Space Centre. https://spacecentre.co.uk/blog-post/the-kessler-syndrome/

Thompson, A. (2020, February 11). *What's next for Solar Orbiter after its historic launch to the Sun.* Space.com. https://www.space.com/solar-orbiter-launched-whats-next.html

Trafalgar. (2020, March 26). *11 incredible sights on Earth that can be seen from Space.* Real Word. https://www.trafalgar.com/real-word/earth-seen-from-space/

Valentine, K. (2017, December 15). *The Amazing Story of the Cold War Space-Egg Race.* Audubon. https://www.audubon.org/news/the-amazing-story-cold-war-space-egg-race

Velcro Companies. (2019, July 16). *HOW ARE VELCRO® BRAND FASTENERS USED IN SPACE?* Velcro. https://www.velcro.com.au/blog/2019/07/how-are-velcro-brand-fasteners-used-in-space/

Vocabulary.com. (2020). *astronaut - Dictionary Definition*. Vocabulary.com. https://www.vocabulary.com/dictionary/astronaut

Wall, M. (2011, March 16). *10 Surprising Facts About NASA's Mercury Probe*. Space.com. https://www.space.com/11147-nasa-mercury-spacecraft-surprising-facts-messenger.html

Wall, M. (2019, April 23). *The Most Extreme Human Spaceflight Records*. Space.com; Space. https://www.space.com/11337-human-spaceflight-records-50th-anniversary.html

Warnock, L. (n.d.). *Sound Suppression System*. Www.nasa.gov. https://www.nasa.gov/mission_pages/shuttle/launch/sound-suppression-system.html

Werries, M. (2016, July 18). *Space Shuttle Tiles*. NASA. https://www.nasa.gov/aeroresearch/resources/artifact-opportunities/space-shuttle-tiles/

Wilson, D. H. (2010, January 27). *Spirit, NASA Martian Exploration Rover, Dies at 6* (Earth Years). Popular Mechanics. https://www.popularmechanics.com/space/moon-mars/a5164/4343770/

Fun Facts Space Trivia 3.0

Pantheon Space Academy

INTRODUCTION

Astronomy is all about discovery, uncovering fascinating facts about the observable Universe. At Pantheon Space Academy, our mission is to challenge your space knowledge, make it fun, let it be informative, and reveal the mysteries of outer space. Welcome to *Fun Facts Space Trivia 3.0!* We dug deeper, traveled further, and uncovered more fascinating facts than ever before.

It's time to put your Space experience to the test, by yourself, or bring your friends and family along for an exciting trip around the Universe. Our book has 165 trivia questions to challenge everybody, from students to novice astronomers and above. We followed the format from previous books 1 and 2. Inside, we ask a question, then receive the answer with over 751 super fun facts on the subject. In this book, you'll find Space facts about the planets, moons, stars, cosmic rays, and that time when astronauts lost $100,000! Stay until the end for a particular thought-provoking chapter on unsolved mysteries.

Grab a notepad to keep score or just read along to learn more. You'll have a fun time testing your knowledge of the extraordinary and ever-expanding Universe. Without further ado, let's gather the crew, suit up, and blast off into the beautiful cosmos!

CHAPTER ONE

OUR PLANETS AND MOONS

Starting the book with a bang, When was the last known meteor strike to hit our moon, Luna?

January 20, 2019. The impact flash, 0.28 seconds, was seen during a total lunar eclipse, aka Blood Moon. Marking the first-ever recording of such an event. A similar observation was made in January 2000, leading some to speculate if the Sagittarids/Capriconids meteor shower could be responsible for both.

Earth and the Moon have been companions for how many years?

Four billion is the current estimate. Scientists have still not confirmed how our moon arrived, but a popular hypothesis is that we received our Luna after a collision between Earth and an object.

Which planet has a similarly cratered surface compared to our Moon?

Mercury, slightly bigger than our moon, has thousands of impact craters on its surface. The giant Caloris Basin impact, caused by a massive object, sent shockwaves through the core, causing hills on the opposite side of the planet! The diameter of Mercury is 3,030 miles.

How many planets have a greenhouse effect?

Three. Earth, Mars, and Venus though they are dramatically different. Earth enjoys the Goldilocks Zone; without it, the temperatures would drop about 90 degrees Fahrenheit, turning Earth into a frozen rock! Mars' atmosphere is too thin to trap heat, and Venus' is too thick and ultimately brings its temperature high enough to melt lead.

True or false: The Gagarin line is the line between Earth's atmosphere and space?

False. The Karman line is the boundary, sixty-two miles from our surface. Not Gagarin line, as in cosmonaut Yuri Gagarin, he was the first person in space and orbited Earth for 108 minutes.

Which planet is the hottest in our Solar System?

Venus, even though it's second from the Sun. The runaway greenhouse effect is primarily thanks to sulfuric acid clouds and carbon dioxide in the atmosphere, trapping the Sun's heat, cranking up the intensity. The same sulfuric acid clouds make Venus astonishingly bright; sunlight bounces off the lighter white clouds and can make this planet easily visible during our daylight hours.

True or false: Mercury, the planet closest to the Sun, has ice water?

True. Scientists deliberated for more than 20 years if the Mercury poles had ice because the north and south poles lie in constant shadows. Eventually, the spaceprobe Messenger would confirm stable ice water in the hollows on the surface of Mercury. Cavities were likely created, not by the atmosphere, but evaporating gases exposed by meteor impacts!

By volume, which Solar System sphere has the most water?

Ganymede. The moon of Jupiter is 46% water, and it happens to be an underground ocean. Earth is fifth on the list; water only credited 0.12% of the volume on the blue marble.

True or false: Europa, a moon of Jupiter, has water?

True! The surface is an icy shell polluted with radiation and gases from Jupiter, believed to have a global ocean under that crust, estimated to be 16% water by volume. Astrophysicists have evidence of water vapor plumes that reach up to 125 miles above the moon's surface. Scientists are eager to explore Europa, saying it is the most likely sphere to harbor life in our Solar System.

Name the planet which has three moons with evidence of water?

Saturn. Titan, the largest moon of Saturn, has the second-largest concentration of water in the Solar System. Enceladus and Dione collected water, but with smaller amounts than Earth. Enceladus touts a surface covered in fresh, clean ice!

Outside of Earth, Which world has liquid on its surface?

Titan is the only sphere known to have liquid on its surface. Surface pressure is fifty times greater than ours, but the atmosphere has clouds, rain, rivers, lakes, and seas of hydrocarbons. Underneath the crust of Titan is a massive ocean, hundreds of feet deep and miles wide!

How many planets have seasons?

All eight! Astrophysicists use the same seasonal names for all planets; summer, fall, winter, and spring. Each planet's seasons are different lengths, mainly attributed to their axis tilt.

Which planet has the shortest season?

Venus. Occurring about every 55-58 days. Earth was the next closest, with 90-93 days separating the seasonal change.

Which planet has the most extended season?

Neptune. With an astonishing 40 years between seasons! Uranus at 20 years is the closest competition for the title. The seasonal changes are more volatile on planets as their distance from the Sun increases.

How many rings does Neptune have?

Five. Ranging from microscopic to some objects reaching the size of a house. Three of Neptune's moons are within the ring structures, leading scientists to believe that the moons hold the rings in a stable orbit.

How many moons does Neptune have?

Thirteen. The most significant and most interesting is Triton. The orbit of Triton is retrograde, meaning it's opposite of the planet's rotation. Astronomers speculate that Neptune grabbed the moon from the Kuiper Belt in the not-so-distant past.

True or false: Triton is the only large moon with a retrograde orbit?

True. Trick question because we said large moon. Five other moons have this orbit pattern, though they are a tenth of the size of Triton. Jupiter claims four of the five, Saturn has one.

Not including Earth, how many planets have active volcanoes?

One. Venus has more than one thousand volcanoes. Astronomers cannot see these through telescopes, but scientists have by using radar.

Name the three moons with active volcanoes?

Triton, Enceladus, and Io. Cryovolcanoes are heated water and gases released into the atmosphere; Triton and Enceladus have these geyser-like events. Io has impressive volcanic sulfur plumes forced high enough into the atmosphere to be seen from far away spacecraft.

Above sea level, what is the name of the first layer of Earth's atmosphere?

Troposphere. The layer continues about six miles above our surface. Planes fly just beneath the invisible borderline to maintain their flight. Earth has a total of five layers of the atmosphere.

What is the fifth and final layer of Earth's atmosphere?

The Exosphere. The layer is home to most satellites and even the ISS. There is no exact line for the start and finish of each layer; the atmosphere just keeps getting thinner and thinner.

True or false: The last detection of the atmosphere is 1,200 miles above Earth?

False. Even closer at 600 miles. The ISS orbits Earth at the height of 200 miles above sea level. The Starlink satellites are 340 miles above the surface.

Which moon has the most distant orbit from its planet?

Neso the moon of Neptune. It travels millions of miles away from the host planet. Neso resistantly was discovered in 2002. Not so far from Neso is Psamathe, and they are suspected to be two halves of a once bulkier moon. Further discovery will be needed to solve if that is true.

In 2017, Perijove 6 spacecraft, seen an arrangement of surface clouds nicknamed?

Jovey McJupiterFace. People online would go on to share the image millions of times! Two white circular clouds we perceived as the eyes and an oval red shape for the mouth. Three weeks later, the emoji-like image was erased by heavy winds and never seen again.

True or false: Luna dust is an out-of-this-world problem?

True! The Apollo missions struggled heavily with the dust and recorded all kinds of issues, such as damaged space suits, clogged mechanisms, instrument interference, causing radiators to overheat. A minor nuisance on Earth but another significant danger in space.

What year did Earth have its first color picture taken while the entire sphere was in the frame?

August 1967 credited to The Department of Defense of Gravity Experiment (DODGE) satellite. The first image from space was in 1946, at the height of 65 miles, from a 35mm camera strapped to a captured German V-2 ballistic missile. That was not in color and not high enough to see the entire globe.

What year did Voyager capture the first image of Saturn?

1980, just 25 days before the spaceprobe would make its closest approach before continuing its journey to Uranus.

When did NASA capture the first image of Venus using a spaceprobe?

On February 5th, 1974, by Mariner 10.

True or false: Pluto's mass is more prominent than our moon, Luna?

False! The Luna diameter of 2,175 miles easily shadows the smaller Pluto diameter of 1,490 miles. Though, Pluto could easily win other comparisons!

How many days does it take our moon to make a complete orbit around Earth?

Twenty-seven. The two are perfectly in sync by rotating at the same speed. The same side of the Moon faces Earth every day!

Which planet has the record for the biggest crater in the Solar System?

Mars - North Pole Basin crater is 6,550 miles wide. Having the record for over fifty years now, but competition is heating up! Astronomers recently have been studying a basin on Ganymede with a diameter of 9,693 miles. Once confirmed to be an impact crater and receiving an official name, there could be a new number 1.

Which meteor "show" was named the most extraordinary meteor shower of all time?

The Leonids of 1833 are the undisputed champion in this category. During the peak of the show, there were possibly 100 thousand meteors per hour. Leaving a lasting impression on those who witnessed, with one viewer writing, "upwards of 100 lay prostrate on the ground... with their hands raised, imploring God to save the world and them. The scene was truly awful; for never did rain fall much thicker than the meteors fell towards the Earth; east, west, north, and south, it was the same."

What hides on the dark side of our moon Luna?

The Aitken Basin, located on the south pole, is one of the largest and oldest impact craters in the Solar System! The impact left a crater measuring 1,600 miles wide and up to five miles deep.

In 1994, Fireworks on Jupiter was a world phenomenon. Why?

The Shoemaker-Levy 9 Comet happened to be discovered a few months before crashing into Jupiter. Astronomers from around the world were filled with anticipation to watch in July. Unknowingly, viewers would receive a 7-day experience as 21 comet fragments bombarded the sphere!

Many moons above, how many are in our Solar System?

One hundred fifty natural satellites and rising! Jupiter and Saturn claim over one hundred of them.

True or false: Mars can have large-scale dust storms that obscure the entire globe?

True. Due to the dry planet, thin atmosphere, and 70 mph wind speeds. The average wind speed is only 20 mph and consistently causes dust devils, but the stronger winds pick up so much dust and light sand, hiding the entire martian surface for weeks at a time.

What year was the first visual of a Jupiter lightning strike?

In 1979, Voyager 1 saw lightning strikes on its flyby. Predicted in 1921 but was not photographed from Earth until 1989. Then in 2016, the Juno mission would seize the opportunity for an in-depth look at these lightning storms. Unlike our planet, Jupiter has three types of lightning.

The Earth and Moon are an example of astronomical bodies making what motion?

Tidal locking aka synchronous rotation. The same side of the Moon is always facing the Earth. Locking is normal behavior for the satellite to lock but not the object with more mass. When the two objects are similar in mass, as is the case for Pluto and Charon, they face each other 24/7.

True or false: More mass equals more gravity?

True! Each planet has a different mass, and your weight would vary on each. The gravitational pull dissipates as you move away from each object. Meaning as you move away from the mass, you get lighter and lighter.

During orbit, what name identifies when a planet reaches its closest point to the Sun?

The perihelion. When the planet reaches its most significant distance from the Sun, it is called the aphelion during orbit. Astronomers make calculations to show how eccentric of an elliptical that the planet has during orbit.

How many planets have an orbit that makes a perfect circle around the sun?

Zero! All orbits have a degree of variance, often called eccentricity. Venus is the least eccentric at .007, and mathematically that is a range of 1.3 million miles. Mercury has the highest variance of the eight planets with a range of 15 million miles of eccentricity.

How many AU, astronomical units, is Neptune from the Sun?

30 AU. The sunlight takes 4 hours to reach the planet compared to Earth, where we receive light in 8 minutes.

How many AU would you travel before reaching the Oort Cloud?

5,000 AU! No man-made object has made it here yet. The Kuiper Belt ends at 55 AU, Pluto is 68 AU, and Voyager 1 at 153 AU. The iconic spaceprobe is traveling 38,000 mph with an expected arrival to the edge of the Oort cloud in 300 years.

Which planet is often referred to as a failed star?

Jupiter. The gas giant mass is not enough to start the fusion necessary for a star. It can be misleading to refer to the planet alongside a star because Jupiter does not have the proper chemical composition to ignite, even if it did have enough mass.

World Record: bulkiest meteor ever found was in which country?

Namibia. The Hoba meteorite is iron-rich, weighs 65 tons, and was discovered in 1920 by a farmer. It is estimated to have struck Earth 80 thousand years ago but oddly shows no impact crater around it. Is that due to its shape, which may have slowed down the entry speed?

What is another name for Aurora Borealis?

The Northern Lights. If you were standing in the southern atmosphere, this phenomenon is called the southern lights, Aurora Australis. Scientists call this space weather.

True or false: Earth is the only planet to experience the Northern Lights.

False. The spectacular light show is not unique to Earth! If you've seen the aurora on Earth, you know how exciting it would be to see them on other planets.

How many of our neighboring worlds have auroras?

Six. All of the outer solar system planets, Jupiter, Saturn, Uranus, and Neptune. Mercury is the only planet to have never been documented with the fantastic light show. Mars and Venus are creating these lights in different ways than the others.

True or false: Luna, our moon has water?

True! In 2008 the Indian mission Chandrayaan-1 discovered trace amounts spread across the surface. Following the path with higher traces led scientists to more significant amounts of ice water at the poles. This water will help astronauts to have longer stays on the lunar surface.

The Earth and planets orbiting the Sun are known to be in what motion?

Heliocentrism. First on record in 300 B.C. by Aristarchus of Samos. His work would attract little attention and the period was not efficient at saving documents. It would be the 16th century before the concept gains traction.

CHAPTER TWO

THE SOLAR SYSTEM

What is the name of the Sun's atmosphere?

Corona. Prominence and solar flares are the loops of gas bursting through the corona. Those are storms that are many times more massive than Earth. Raining lava is one way to describe a solar prominence.

When was the last observed solar flare massive enough to affect Earth?

October 6th, 2020. Had the CME, coronal mass ejection, happened a few days earlier, the trajectory would have been a direct hit on Earth. More minor flares are regular and cause the auroras here on Earth.

Which year was the last major CME to hit Earth?

1859. The Carrington Event, 161 years ago, caused an electric surge. Telegraph systems across Europe stopped working, and the upsurge in electricity caused some buildings to set on fire. A solar storm of similar size in modern times could damage the electric grid, televisions, radio, and the internet.

What type of celestial body is Hale-Bopp?

Long-period comet flaunting 37 miles of ice and rock, referred to as the Great Comet of 1997. Astronomers and citizens gazed for 18 months during 1996 and 97 with a trajectory to arrive back within view around 3731.

Pioneer 11 was the first spaceprobe to visit which planet?

Saturn. The visit was brief and classified as a flyby. Pioneer 11 is the fourth spaceprobe beyond our Solar System carrying a message, should it ever be intercepted by intelligent life.

Which asteroid was first to be orbited and landed on by spacecraft?

433 Eros, the Greek god of love, carries the title of the first. NEAR landed on Eros on Valentine's Day in the year 2000. It was love at first sight! A spectacular start to the new millennium.

What makes the orbit of Pluto unique compared to our eight planets?

Its incline of 17 degrees. Mercury, the most eccentric orbit of the eight planets, is just 7 degrees inclination.

True or false: at times, Pluto is closer to the Sun than Neptune?

True. When is the next time this will happen? 2227. This last occurred from 1979 through 1999. Pluto is one of the most patient spheres in the Solar System.

What do Pluto and Uranus have in common?

They both rotate on their sides. The Plutonian axis is tilted about 120 degrees.

Scientists discovered that the Plutonian mountains are of what substance?

Ice! They stand about a mile high. Pluto is an exciting sphere for astronomers. Waiting on future discoveries has us on the edge of our seats.

True or false: Asteroids have no water inside?

False. Scientists at the Ritsumeikan University of Japan studied a 4.6 million-year-old asteroid containing a small amount of liquid containing at least fifteen percent carbon dioxide. Asteroid water will come in handy when travelers are thirsty during space travel.

Which arm of the Milky Way is our Solar System resting?

Orion Cygnus Arm is our very own spiral arm! Thanks to this small quiet arm of the galaxy, we enjoy a relatively stable Solar System. Gladly not in the Sagittarius and Perseus arms, which have a very violent presence.

OSIRIS-REx gathered a 60-gram rock and mineral sample from which asteroid?

Bennu. The spacecraft launched, in 2016, towards the destination of a carbonaceous near-Earth asteroid and will return with the sample in 2023.

Which asteroid's mass makes up nine percent of the entire asteroid belt?

Vesta, the second-largest body in the asteroid belt. Thought to be a protoplanet before its growth was interrupted.

Name the celestial object with the most mass in the asteroid belt?

Dwarf planet Ceres, discovered by Giuseppe Piazzi, is 25 percent of the asteroid belt's mass. The name is from the Roman goddess of corn and harvest, also translated to cereal.

Which space rock spins so fast that it is now an egg shape?

Haumea dwarf planet. Three hundred and eighty-five-mile radius, 1/14 the size of Earth, with a 4 hour day makes this oval-shaped rock one of the fastest spinning objects in the Solar System.

Comets are being produced by which two regions of the Solar System?

The Kuiper Belt and The Oort Cloud. These comets, upon discovery, will be labeled (P) periodic, (C) non-periodic, (X) no orbit, and (D) lost. Losing comets will happen for a few reasons. One is that of non-gravitational forces, for example, emissions of jet gases from the nucleus.

Are there more or less than 3,000 comets known to exist in our Solar System?

More! While it is always hard to get an exact count in outer space. Astronomers have been documenting comets since ancient times. The number fluctuates as new comets appear, and old ones collide with other celestial objects and end their time on the list.

Universe Record: How long was the most extended comet tail?

Six hundred twenty-one million miles by comet 153P/Ikeya-Zhang. Imagine if this tail were visible to humans from Earth; the ion tail would have stretched over half the sky! The previous comet holding the record was Hyakutake's tail and measured half the length. In competition, that's called "blown away"!

Which comet streaked through our Solar System in July 2020?

NEOWISE. Astronomers and stargazers eagerly waited for the comet to make its closest approach to Earth. The comet is 3 miles wide, and some estimates say it holds enough water to fill 13 million Olympic-sized pools. The last time Earth saw the comet, 6800 years ago; not even Stonehenge existed. Our planet has seen a lot of notable changes since then!

How many asteroid orbits cross paths with Earth's orbit?

Three. They are Apollo, Eros, and Adonis. They have a lot of eyes watching them!

Which asteroid of Jupiter has an orbit that also goes around the Sun?

Hidalgo. The asteroid is not a threat to Earth at this time. Speculated to be a failed comet and captured by Jupiter. The eccentric orbit around Jupiter includes rounding the Sun at approximately 2 AU.

Closest comet to pass by the Earth?

Comet P/SOHO 5 passed Earth on June 12th, 1999. The tiny comet almost moved through the Solar System unseen. The little tail was intermittent at best. It was about five times the distance from Earth to the Moon. Comet Lexell held the previous record in 1770.

Name the first (documented) object from another star to visit our Solar System?

The Oumuamua comet. She was first discovered on October 19, 2017, by the University of Hawaii. Its name means "a messenger from afar arriving first." An impressive interstellar visitor measuring ¼ of a mile in length and only a tenth as broad, a proper shape is a cylindrical capsule, similar to a hotdog.

True or false: the Kuiper Belt abruptly ends at 50 AU?

False, for now. The belt was predicted in 1950 by Gerard Kuiper and confirmed in 1968. The running theory is that the icy rocks in the belt would slowly dissipate as you move through. But with more research, astronomers now propose the Kuiper Cliff as a sharp drop off of icy objects. For this theory to be correct, there would have to be a Planet 9 and Planet X pushing the smaller celestial objects into a thicker formation to create the cliff.

CHAPTER THREE

BEYOND OUR SOLAR SYSTEM

True or false: A satellite galaxy is a smaller galaxy that orbits a more massive galaxy?

True. Satellite galaxies pulled by gravity to a much more massive universe. Similar to how the Sun's gravitational pull trapped the planets. The Milky Way currently has fifty satellite galaxies in orbit.

True or false: The Large Magellanic Cloud galaxy is our closest galactic neighbor?

False! Using its massive clouds, LMC is rich in star formation and was once known as the nearest. Further discoveries have now placed it third. Sagittarius Dwarf Spheroidal galaxy discovered in 1994 had the title until new evidence makes Canis Major Dwarf galaxy the closest.

What is the natural state of the universe?

Darkness. "Let there be light" is an afterthought to a problem, constant darkness. We wouldn't see anything without those giant radiating balls of fire. Wherever light travels, it finds that darkness was already waiting for it.

What percentage of stars have a planetary system like our Sun?

Eighty percent of stars have planets in orbit. Some estimates go as high as one hundred percent. Every star you see at night could be somebody else's Sun.

In the observable universe, astronomers calculated approximately how many galaxies?

Two trillion! Amazing how life has survived, evolved, and thrived in such a complex world. With intelligence and impressive technological advancements that humans created on Earth. It leads one to ask, Are we alone?

What effect allows astronomers to see black holes?

Gravitational lensing, discussed in Albert Einstein's general theory of relativity. The light from a distant source bends around an anchored black hole sitting between the light source and the observer.

True or false: Gravitational lensing only appears around black holes?

False. Recently astronomers pointed to a cluster of galaxies to observe the gravitational lens effect. Thought to be caused by dark matter bending light. Though this may be our first visual evidence of the mysterious substance, it is safe to say that physicists will need more time investigating.

Over 100 thousand stars bound in a group by gravity is called what?

Globular clusters and can contain up to one million stars! Ten billion years old and some of the oldest stars to exist. The Milky Way is known to hold 150 globular clusters.

Which star cluster is known for its striking blue reflection?

The Pleiades or Seven Sisters star cluster. The dusty nebula nearby is highlighted by the star cluster to make a remarkable view. Four hundred light-years away, scientists have singled out one young Pleiades star with a terrestrial planetary zone similar to ours.

What do astronomers call a galaxy that creates hundreds of stars per year?

A Starburst galaxy. Typical star formation in a universe is one or two per year. Suppose a galaxy holds enough gas, the production rate ramps up to hundreds per year. A trigger is necessary for the universe to enter starburst mode, collisions! Lots of them.

What cosmic structure happens in front of the fastest, most massive stars?

A bow shock. Astrophysicists can tell a lot about the conditions around the star and in space when they research a bow shock, including magnetic fields, particles flowing off of a star, solar winds, gas, and dust filling the distance between the stars. Our Sun's bow shock is nearly invisible because of how slowly it moves.

What do astronomers call a molecular cloud that is in the process of forming new stars?

Stellar nursery or star-forming region.

Which nebula has very young stars less than one million years old?

NGC 1333, not a fancy name but could be similar to the chaotic birth of our Sun. The nebula has hundreds of newborn stars surrounded by glowing gases and thick stardust with red and bluish hues. You can view this nebula in the constellation Perseus.

What did astronomers name the closest stellar nursery to Earth?

The Orion Nebula is about 1,500 light-years away. One of the most observed and photographed objects in the sky. Deep within the nebula are young, massive, and luminous stars.

A quasar gives off giant amounts of what?

Electromagnetic radiation. The energy is released when gas from the disk falls into the black hole. Powerful quasars are luminous enough to outshine an entire galaxy.

Which type of nebula absorbs a hot star's energy, causing its gases to glow?

An emission nebula. The color varies according to ionization and the chemicals inside the nebula. Two examples are The star-lit Eagle Nebula and the Pacman Nebula.

Universe Record: Name the most massive star known to exist?

Star R136a1 is 265 times the mass of our Sun and is almost double the size of the next biggest star. Astronomers are still pondering how this is possible, but the Large Magellanic Clouds are known for massive stars. When you're ready to see this star look inside the supercluster of the Tarantula Nebula

Universe Record: Name the least massive star?

OTS 44 is a Universe featherweight at 1.5% the mass of our Sun, equal to 15 Jupiters. The brown dwarf is magenta at 550 light-years away in the southern sky constellation of Chamaeleon.

Visible to the naked eye, which star has the largest diameter?

VY Canis Majoris. Its estimated radius is two thousand times that of our Sun. The star is 3,900 light-years away, but if moved into our Solar System, the edge boundary would be about the orbit of Saturn. You can see this star in the constellation of Canis Majoris.

Universe Record: Which globular cluster holds the most stars?

Omega Centauri holds about 10 million stars, far beyond the next closest cluster. The stars within are older, redder, and less massive than our Sun. The group is visible at night inside the constellation Centaurus, but it's inside our Milky Way Galaxy!

Universe Record: The most massive nearby galaxy?

The breathtaking elliptical M87. It contains 800 billion Suns' worth of mass. M87 includes a supermassive blackhole, several trillion stars, and one hundred fifty thousand globular clusters. In comparison, the Milky Way Galaxy has 300 billion stars and one hundred fifty globular clusters.

What is the proper name for the smallest of galaxies?

A Dwarf galaxy. Our Universe is rich with dwarf galaxies but typically only found as a cosmic companion to a more sizable galaxy because of their size and low luminosity.

Which is the nearest galaxy to our Milky Way?

The Canis Major, dwarf galaxy, in the constellation of Canis Major, received the honors in 2003. Discovery came when astronomers investigated a string of stars behind the universe and discovered three rings around the Monoceros Rings galaxy. When multiple globular clusters came into view, we officially had a new, very close galaxy!

Universe Record: The tiniest galaxy is home to how many stars?

One thousand stars! Segue 2, a dwarf galaxy, has far fewer stars than any previous records. The universe is fragile, bordering on classifying it as a star cluster if the glue of dark matter doesn't hold. Cosmologist James Bullock said, "finding a galaxy as tiny as Segue 2 is like discovering an elephant smaller than a mouse".

Name the point at which nothing can escape a black hole, not even light?

The event horizon. Black holes only consume matter that crosses this invisible line. One example of this is Swift J1357.2, a star that orbits a black hole every 2.8 hours, the shortest orbital period known to man but has yet to penetrate the point of no return.

What is the name for matter build up around the black hole?

Accretion disk. The black holes eyes are bigger than its stomach. Build-up is from collecting matter that the black hole has not munched up. The black hole's lack of appetite allows a star such as Swift J1357.2 to orbit without being instantly torn apart.

When was the first planet found that orbits a star and not our Sun?

In January 1995 by an astronomy graduate student, Didier Queloz, in southern France. He was innocently in the library writing code and checking data that he previously captured when data suggested a wobbling star. 51 Pegasi had a stellar jostling and was later confirmed to have a planet! That was a special day for Queloz but also for astronomers.

Name the two clusters relatively young and rest in the Perseus constellation?

The Double cluster or independently named NGC 869 and NGC 884. A favorite for new astronomers because it's easy to find in the sky and peaks in August. It covers an area twice the size of a full moon and is very bright. The duo each has three to four hundred stars and mostly of blue-white supergiants!

After the rings of Saturn, name the next most famous band from a planetary nebula?

The Ring Nebula (M57). Found to be the remnants of a dying star. It's a cosmic object that leaves many speechless as the color and shape hypnotize you. It's as if you're looking in a window to another world. Relatively close at about 2,000 light-years away. Roadtrip?

Revolving around a common center, what do we call two stars linked by mutual gravity?

Binary star system. The method of detection determines how the binary stars are classified. The most common are spectroscopic binaries that astronomers detect by Doppler shifts in their spectral lines.

Which nebula holds the most prominent star in our Milky Way galaxy?

NGC 3603 nebula birthed a Wolf-Rayet star in the Carina spiral arm. Wolf-Rayets can be up to 120 times the mass of our Sun. It is 22,000 light-years away and known as one of the most luminous and compact clusters in the galaxy.

The Large Magellanic Cloud gave astronomers their first view of what rare cosmic activity?

A supernova! The exploding star was first noticed on February 24, 1987. That is why the name became Supernova 1987A.

What cosmic object was Edmond Halley referring to when he said, "This is but a little patch, but it shows itself to the naked eye when the sky is serene and the Moon absent."?

The Great Globular Cluster. Located in the constellation of Hercules, once known as M13. The cluster core has 100 stars within a three-light-year cube. In comparison, the Sun's one neighboring star is four light-years away.

One thousand five hundred light-years from Earth, what is "The Unicorn"?

A black hole. We understand black holes to be large dominating forces in space, but this stellar beauty is tiny! One of the smallest black holes known to man. Unicorn received the name because of how rare its low mass is at just three times the mass of our Sun. Ohio State University found the Unicorn when a nearby star started ever so slightly to distort.

CHAPTER FOUR

ALL ABOUT ASTRONOMY

What is the IDA?

International Dark-Sky Association. The movement started in 1988 to protect our nighttime views. More than 75 US National Parks have been certified nocturnal, protected, starry skies. Over 51 countries have made efforts to minimize light pollution. Thousands of astronomers and millions of the population find these areas to be a treasure chest of stars!

How many satellites orbit Earth as of April 2021?

7,389. Satellite launching had a 28% increase from the previous year, setting a single-year record of 1,283 orbit additions. Starlink has eighteen hundred in their constellation with plans to have more than 4,000. It's hard to estimate how many are not working, but some estimates say just under half of the 7,389 are floating junk now. Eventually burning up in Earth's atmosphere.

When was the last time that all humans were living on Earth together?

On October 31st, 2000, three astronauts launched from Baikonur cosmodrome headed to the international space station (ISS). Spacecraft arrived two days later on November 2nd, 2000, unintentionally starting a chain of constant human presence on the ISS more than two decades later.

How many humans have visited the space station?

Two hundred forty-two individuals from 19 countries have participated. A typical crew aboard the ISS is six at a time. However, this has been up to 12 astronauts simultaneously. While aboard, astronauts admire Sixteen sunsets and sixteen sunrises every 24 hours while traveling at 5 miles per second around Earth.

Which astronaut set the world record for "most total time living and working in space"?

Peggy Whitson. She has tallied 665 days in space! Two hundred eighty-nine of those days were consecutive. The most extended individual duration in space was by Valeri Polyakov at 437 days. Astronauts have to use a water recovery system aboard spacecraft to cut down dependency on the amount of liquid cargo. Increasing their ability for longer durations off Earth!

How many space probes have exited our Solar System to be interstellar probes?

Two. Both Voyager 1 and 2 have crossed the heliosphere and continue to transmit data to NASA. Scientists have shown a decrease in heliosphere particles and a considerable increase in galactic cosmic rays upon analyzing the data! Crossing the line was a 35 and 41-year journey for the duo.

True or false: Pioneer 10 and 11 will enter interstellar space as ghost ships?

True. Unfortunately, both probes stopped functioning during their exploration. Pioneer 11 is headed towards the galactic center and sadly will not be sending us any pictures.

NASA's Goddard Space Flight Center is in which city and state?

Greenbelt, Maryland. This is the home to our nation's largest organization of scientists, engineers, and technologists who build new technology spacecraft and instruments to study planets, the Solar System, and beyond. See their website for virtual tours - https://www.nasa.gov/nasa-at-home-virtual-tours-and-augmented-reality.

True or false: The TBIRD mission will study the Phoenix Spots on the Sun?

False! NASA 2021 will be testing laser communications as a faster, more efficient way to downlink from outer space. Bright lasers won't fill the night sky since the technology uses infrared light similar to a TV remote. The transfer goal is 200 gigabits per second, equaling 250 movies that are each 120 minutes long.

What technique allows spacecraft to swing from one planet to another?

The gravity assist technique. First tried during a rare planetary alignment that only occurs every 176 years. Voyager 1 and Voyager 2 used this assistance perfectly, which allowed them to travel further and faster without heavy propulsion systems.

The James Webb Space Telescope is orbiting Earth at what distance?

One million miles! The size of a tennis court and known as the successor to the Hubble Telescope, though Hubble orbits very close to home. The JWST will be a significant upgrade with instruments and a 21-foot mirror for capturing light, ultimately allowing NASA to look further into space than ever before.

What was the name of a famous Voyager 1 photo, 4 billion miles away from Earth?

The Pale Blue Dot captured Earth on Valentine's Day 1990. It continues to hold the record for the most distant picture ever taken of Earth. Voyager 1 was beyond Neptune at the time, and because of the distance and cold, the file took months for NASA to download.

Astronaut, Charlie Duke, left what on the moon?

Family portrait. During the Apollo 16 mission, he intentionally left the picture on the surface of Luna in 1972, but the image will be completely white now due to the Sun's radiation.

Astronomers who study the structure of the Universe are called?

Cosmologists or extragalactic astronomers. Specialized and trained to study creation, evolution, possible futures of the universe, and the galaxies.

True or false: More than 300 robots explored outer space beyond Earth?

True. NASA would call this "leaving the cradle." That's a mountain of discoveries made, and with the private sector creating competition, the robot number will grow exponentially within the decade.

What mission was the reason to repurpose the Kepler Space Telescope?

Planet-hunting. For two months, Scientists used the data collected to identify rogue and exoplanets. Using a technique of gravitational microlensing. After 27 hits and four believed to be planets, the size of Earth, the mission would end, and the telescope officially retired.

How many years were in between the first winged flight on Earth and astronauts landing on the moon?

Sixty-six years! That is human ingenuity at its finest. The two astronauts walked on the moon for 3 hours, collecting samples, planting a US flag, and leaving a sign that said, "Here men from the planet Earth first set foot upon the moon July 1969, A.D. We came in peace for all mankind."

Name the NASA designated area for astronaut's final preparations before launch?

The White Room. Located inside the shuttle and intentionally all white. A particular room ensures that dust, dirt, and stray hairs would not stay in the shuttle. The final touches in this room were for checking parachutes and putting on helmets. Now crews refer to the end of the Crew Access Arm as the white room.

Astronauts on the ISS are researching the growth of protein crystals for what purpose?

Medical advancement. Gravity interferes on Earth, but the proteins grow quicker in microgravity, allowing scientists to speed up testing times, focusing on using drugs to target and treat diseases. This process did help make a beneficial insulin treatment.

True or False: Astronaut lost control of a toolbag during a spacewalk on the International Space Station?

True! On November 18th, 2008, A grease gun leaking and the bag being unsecured was enough to propel the load away from the astronaut. Hundreds of amateur astronomers tracked the 30-pound toolkit, backpack size, valued at 100,000 dollars. Eight months after silently orbiting Earth, the bag eventually burned up in our atmosphere.

Has an astronaut ever died while in outer space?

Yes. While most astronaut deaths have been on earth during the rigorous training period, three astronauts met their demise when exposed to the vacuum of space. Known as the Soyuz 11 disaster, on June 29th, 1971, right after the space crew spent a record-breaking three weeks aboard the Salyut 1.

During the launch, humans experience how much g-force?

3gs. That is equivalent to 3 times the gravity that we feel here on Earth. During reentry into Earth's atmosphere, the astronauts can experience 3gs for up to 15 minutes. With proper training, the human body can safely perform.

How much g-force can an astronaut withstand before damaging their body?

9gs. This is nine times the gravity that our bodies will withstand. Once the body is under such pressure, damage starts occurring, and you blackout as blood cannot reach the brain.

True or false: Voyager 1 will send us hundreds of pictures when it arrives at the Oort Cloud?

False. Sadly the power supply will be depleted long before arrival. At this distance from the Sun, it cannot recharge from solar. NASA did a tiny test recently, and Voyager 1 still has power for now. You can follow the satellite at: https://voyager.jpl.nasa.gov/mission/status/.

Which country has the world's largest radio satellite?

China. The Aperture Spherical Telescope (FAST) became operational in 2016, located southwest of Beijing. The dish size is about 30 football fields; if there is intelligent life, scientists say that this satellite will find it 5 to 10 times faster than current equipment.

The first journey to step on the moon, 1969, had how many critical checkpoints?

Twelve. From launch to landing, NASA planned every inch of the intense mission to put the first man on the moon. Stage 1 is the launch, then the third drop off (stage 5) the astronauts were on moon-bound, stage 7 lunar module landed on the Moon, stage 10 service module discarded, and they were on course for a return to Earth!

While on the moon, Armstrong and Aldrin left what two important communication instruments?

The seismograph and a Laser Ranging Retroreflector will record any "moonquakes" and the other to measure the precise distance from Earth to Luna.

What was the name of the first rover on Mars?

Sojourner. It landed in 1997 but is no easy task because scientists cannot do any controls in real-time and leave the landing up to the programmed computers. This mission did not gain much amazing science but paved the way to benefit from future exploration.

Greeks began the scientific study of the Universe after being influenced by what region?

Mesopotamia in Greek means "between two rivers." This was an advanced civilization in B.C. times. One of the first ideas proposed by the ancient Greeks was that the Sun, Moon, and planets revolved around Earth.

Heliocentrism wouldn't gain traction until which renaissance mathematician and astronomer presented their findings?

Nicolaus Copernicus. His revolutionary way would lead to what is now known as the Copernican Revolution. Leading the way for heliocentrism. In the next century, after the invention of the telescope, Galileo would later confirm the findings. Elliptical orbits introduction soon followed.

Which German astronomer introduced elliptical orbits?

Johannes Kepler. In the 17th century, he was a renowned mathematician, astronomer, and astrologer—best known for his three laws of planetary motion.

Which famous astronomer discovered Uranus?

Frederick William Herschel. The 18th-century astronomer became famous overnight upon the confirmation of the planet in 1781. Following the fame, King George III appointed Herschel Court Astronomer. This would lead to grants and the construction of new telescopes.

William Herschel would then pioneer which type of astronomy for stellar spectra?

Astronomical Spectrophotometry. A breakthrough for studying light in outer space, which would lead to another discovery of infrared radiation. Herschel discovered thousands of nebulae, four moons and improved upon Mars's rotation period.

NASA renamed the Voyager mission to VIM. Why?

The original Voyager Planetary Mission was complete. Voyager Interstellar Mission (VIM) is the next best option for scientists to send spacecraft past the heliosphere, unsure when we would have the chance again. They were right; two other probes have since lost all functions before reaching the distance.

Who established classical mechanics and firmly linked astronomy and physics together?

Isaac Newton. Born on Christmas Day, 1642 would become widely known as one of the greatest mathematicians and most influential scientists of all time. The book of mass scientific appeal is The Mathematical Principles of Natural Philosophy.

Under a dark sky, the naked eye can see how many stars?

About 2,000. These are the stars that astronomers used to draw constellations in the sky.

Name the coordinate system used by astronomers to describe the position of an object or star?

The altazimuth coordinate system or horizon coordinate system. Earth observers use coordinates of altitude and azimuth. Azimuth is the celestial horizon from the observer's south point.

What is it called when you use a few stars from a constellation to make a new constellation?

An Asterism. Civilizations can practice asterisms locally without being recognized intentionally. Before humans kept better records and had quick communication, there were many constellation variances. Of course, today, we have 88 internationally recognized and documented star patterns.

In 3,000 B.C. Egyptians built the pyramids to align with which asterism in the night sky?

Orion's Belt is the asterism in the constellation Orion, known as Osiris, God of the Dead. Theories remain that the Egyptians carved out the Giza pyramids to align with the three stars on Osiris's belt.

The Mayan Pyramid windows align to reflect the rising and setting of which planet?

Venus. Remarkably these ancient civilizations had an excellent grasp of astronomy and could be independent of each other. Without proper documents, we can only speculate the intended use of the windows, maybe Venus, or could have been the equinox. Will we ever know with certainty?

Which man-made object orbits Earth and is the size of a football field?

The ISS. The space station is one yard short of being a full-sized field, including the end zones. Over time additional pieces would lead to six bedrooms, two baths, a gym, and a 360-degree bay window for the most spectacular views. Time from launch to arrival, a spaceship can arrive in 4 hours. There are eight docking stations available!

After a spacecraft mission is complete, they move into what new category?

The mission of opportunity. If the vehicle has more fuel, NASA will assign the craft another mission to continue further investigations.

Which spaceprobe flew by Neptune and discovered five moons, four rings, and a "Great Dark Spot"?

Voyager 2 in 1989. Weighing almost 1600 pounds and fitted with eleven scientific instruments. It continues to be the only probe to study the Solar System's four giants up close. Safe to say that both Voyagers are rock stars!

True or false: NASA created a firework, in outer space, by shooting a comet?

True! On July 4th, 2005, the impactor was placed in the trajectory of comet P/Tempel 1. The impact happened at 23,000 mph, generating an explosion equivalent to 4.7 tons of TNT, and made a 490-foot wide crater. The Deep Impact mission was a success.

What is the next gold rush?

Asteroid mining. It is estimated to be worth over 65 quintillion bucks, that's 18 zeros, in the asteroid belt between Mars and Jupiter! Based on element and mineral content, one small asteroid can be worth trillions of dollars alone. Instead of being called 49ers, they will be Astro9ers!

FINAL WORDS

We are grateful to have shared our passion for outer space and all its amazing facts with you. Fun Facts Space Trivia 3.0 was our most challenging in the series yet; it has been a joy exploring with you. It is always our mission to make space exciting and shareable for the whole family. Please consider leaving us a review; let us know how much you enjoyed the galactic trivia and World Records of the Universe.

Don't leave yet! We have installed a bonus chapter of unsolved mysteries. We ask questions that will lead to more questions. The Universe always keeps some secrets. Explore our ten-question bonus, intended to provoke thought and imagination. Let's get right to it!

BONUS ROUND

UNSOLVED MYSTERIES

Star KIC 8462852 fluctuates by up to 20%. Is this a sign of an advanced civilization?

Researchers have gone through the possibilities of instrument errors, dust clouds, variable stars, and planetary transit. Still, none of this data adds up, leaving scientists to speculate that something is afoot. Could this be the infamous Dyson Sphere, a large structure built around the star to harness its energy? That's what an advanced civilization would do when exploring space for long periods. The 20% swings in brightness could be from power amazingly pulled out of the star! Did this come right out of a Sci-Fi novel? Regardless, the star is too far away to see up-close what is happening for sure.

What is the most massive structure in the known Universe?

The Large Quasar Group. This deep-space area is more than a billion light-years across, and with 73 active quasars on display, it is mind-boggling. Scientists use the cosmological principle to make sense of our galactic neighborhood. It means that wherever we look in the Universe, it should look the same on a macroscale, but this quasar group is far from similar. The megastructure has left scientists puzzled and reminded us how the Universe can always prove old theories wrong.

What is uniquely complicated about the Eridanus Supervoid?

The area of space is empty of everything. It's a billion light-years across, found in the Eridanus constellation. We always think of space as large swaths of openness with objects, gases, and phenomena with giant distances between one another. This void, however, is pristine, not a single spec of matter, not even dark matter! With the advancement of quantum physics, we know that space is never empty. It is leading scientists to make some hypotheses about the area being a parallel Universe. Could this be a gateway to another dimension or the elusive wormhole? "Time" will tell.

The Great Attractor, is it more potent than Dark Energy?

When astronomers observed the Norma Supercluster, they noticed an odd pulling happening to stars and nearby galaxies. This gravitational tugging, specific to this region, has caused galaxies to speed towards it at 200,000 mph. Could this be another link to Dark Energy and the expansion, or have scientists been missing a giant clue right in front of them? Either way, the discovery is a terrific find and will help future equations.

A massive black hole as bright as 420 trillion Suns?

We thought black holes couldn't get any more massive because that would defy the cosmological principle. Scientists can't know everything, and that's what makes the cosmos exciting. J0100+2802 is the most massive and brightest black hole ever found. Leaving astronomers to scratch their heads on this one because early Universe objects should not get this size. SDSS J0100+2802 tops the list as the most mysterious black hole in existence.

Galaxies without dark matter, is this possible?

We have all been paying attention—dark matter, the mysterious substance that is the glue in all of outer space. Even when we see outer space with no visible matter, we can be sure that dark matter is there. That was until researchers found multiple galaxies void of dark matter. If dark matter is missing, how can these galaxies stay together? This is but another question that our Universe proposes to seduce scientists to learn more continually.

Ultra-high-energy Gamma Rays come from where?

In 2021 a burst of energy zipped through the Milky Way. Was this a close call for Earth? The researchers on the Tibetan Plateau used sophisticated instruments to find this intense burst. These cosmic rays come from an unknown source. Gamma rays frequently come from exploding stars and black holes, but these specific gamma rays came from an unknown source. When these particles come into contact with cosmic dust, the result is nuclear reactions. The reaction, in this case, was 957 teraelectronvolts (TeV). The highest scientists ever researched was 6.5 TeV. Leaving us to wonder if Earth could be in the path of the following sequence of gamma rays.

What hellish planet, 500 light-years from Earth, holds two records?

TOI-1431b. Physicists discovered a scorching planet at the University of Southern Queensland. The first record, this planet vaporizes metal with a star size surface level of 4,892 degrees Fahrenheit, making this the hottest planet known to man. The size is astonishing at twice the mass of Jupiter, but this planet has two records, and it's not because of its size. The second record is on the planet's nightside as the second hottest temperature recorded from a sphere at 4,172 degrees! No explanation has proven how a world would heat up to this temperature. Safe to say that aliens will not be coming from this planet.

Did the "Space Roar" come from our very own galaxy?

It was 2006 when scientists sent a weather balloon with special instruments 23 miles into the atmosphere. On a separate mission to pick up radio waves from heat sources, but in return, they heard a loud unexpected roar six times anything they've listened to before. The diffuse signal was coming from all directions, making space one constant sound. We all know that sound does not happen in outer space's vacuum, right? The sound has scientists baffled, and a great debate erupted if the Milky Way created the signal or is it from somewhere else? We have yet to trace the origins of the sound, but it is there and constant.

Alien sightings are through the roof. What has caused the spike?

2023 is on track to be the highest number of reported UFO sightings in history. Gradually increasing since 2014, and in 2020 these sightings were over 8,000! Many sightings do get resolved. Lighting, airplanes, and even the SpaceX satellite launch had reports of people seeing dozens of lights in a perfect trail moving across the night sky. Last year over 2,000 were classified unidentified. We didn't want to compare the missing persons' stats, so instead, we have composed a list of states with the most sightings for you to visit and test your luck for an alien abduction! These ten states have the most reported UFO sightings: Idaho, Montana, New Hampshire, Maine, New Mexico, Vermont, Wyoming, Hawaii, Washington, and Connecticut. If you don't see any aliens, they are still great places to view the stars!

That's the end of our mysteries today! We skated the line from fact to fiction, Karman line to the heliosphere, galactic cosmic rays to UFOs. That was a thrill! Thank you for riding along. Please consider leaving us a review. We will gratefully appreciate you taking the time. If you can't get enough of outer space, just like us, join our FB group: www.facebook.com/groups/pantheonspace and you can download our free eBook, 13 Constellations Visible To The Naked Eye, at www.pantheonspace.com.

Ad Astra!

RESOURCES

Chapter One: Planets and Moons

- *All About Jupiter | NASA Space Place – NASA Science for Kids*. (2021, June 2). NASA Space Place. https://spaceplace.nasa.gov/all-about-jupiter/en/

- Carter, J. (2021, May 18). *How Scientists Used The Last 'Blood Moon' To Measure The Moon, Watch A Meteorite Strike And See Earth As An 'Alien Planet.'* Forbes. https://www.forbes.com/sites/jamiecartereurope/2021/05/15/how-scientists-used-the-last-blood-moon-to-measure-the-moon-watch-a-meteorite-strike-and-see-earth-as-an-alien-planet/

- Center, A. M. S. F. (2017, August 3). *The Greatest Meteor Show of All Time – Watch the Skies*. NASA. https://blogs.nasa.gov/Watch_the_Skies/2017/08/03/the-greatest-meteor-show-of-all-time/

- *Could Jupiter Become a Star?* (2020, January 29). ThoughtCo. https://www.thoughtco.com/could-jupiter-become-a-star-4136163

- Deziel, C. (2019, March 2). *Saturn's Temperature Ranges*. Sciencing. https://sciencing.com/saturns-temperature-ranges-7704.html

- *Greenhouse effect on other planets - Energy Education.* (2020, January 31). Energy Education. https://energyeducation.ca/encyclopedia/Greenhouse_effec t_on_other_planets

- Hoang, B. C. (2019, August 15). *10 Things, Dec. 4: Awe-Inspiring Jupiter Images Not to Be Missed.* NASA Solar System Exploration. https://solarsystem.nasa.gov/news/266/10-things-dec-4-awe-inspiring-jupiter-images-not-to-be-missed/

- *Hoba: The World's Largest Meteorite.* (2006). Geology.Com. https://geology.com/records/largest-meteorite/

- *How many astronauts have died in space?* (2019, October 7). Astronomy.Com. https://astronomy.com/news/2019/10/how-many-astronauts-have-died-in-space

- Howell, E. (2018, January 24). *Shoemaker-Levy 9: Comet's Impact Left Its Mark on Jupiter.* Space.Com. https://www.space.com/19855-shoemaker-levy-9.html

- *In Depth | Earth's Moon –.* (2019, December 19). NASA Solar System Exploration. https://solarsystem.nasa.gov/moons/earths-moon/in-depth/

- *In Depth | Neptune –.* (2019, December 19). NASA Solar System Exploration. https://solarsystem.nasa.gov/planets/neptune/in-depth/

- *In Depth | Neso –.* (2019, December 19). NASA Solar System Exploration. https://solarsystem.nasa.gov/moons/neptune-moons/neso/in-depth/

- *NASA - Seasons on Other Planets.* (2004, July 22). NASA. https://www.nasa.gov/audience/forstudents/k-4/home/F_Planet_Seasons.html

- *NASA's Hubble Spots Possible Water Plumes Erupting on Europa.* (2016, September 26). NASA. https://www.nasa.gov/press-release/nasa-s-hubble-spots-possible-water-plumes-erupting-on-jupiters-moon-europa/

- *Our Solar System.* (2021, February 17). NASA Solar System Exploration. https://solarsystem.nasa.gov/solar-system/our-solar-system/overview/

- Redd, N. T. (2016, August 31). *How Big is Mercury?* Space.Com. https://www.space.com/18647-how-big-is-mercury.html

- Roper, J. E. (2019, March 2). *What Are Some Interesting or Unique Features of Neptune?* Sciencing. https://sciencing.com/interesting-unique-features-neptune-5995009.html

- Rossignol, D. (2018, September 7). *NASA Says Other Planets Have Northern Lights Too.* Nerdist. https://nerdist.com/article/nasa-northern-lights-other-planets/

- Sands, K. (2021, June 10). *Dust: An Out-of-This World Problem*. NASA. https://www.nasa.gov/feature/glenn/2021/dust-an-out-of-this-world-problem/

- *Space Volcanoes! | NASA Space Place – NASA Science for Kids*. (2021, February 11). NASA. https://spaceplace.nasa.gov/volcanoes/en/#io

- Team, H. I. W. (2017, May 2). *Do other planets have auroras?* How It Works. https://www.howitworksdaily.com/do-other-planets-have-auroras/

- *Titan - Overview*. (2019, June 27). NASA Solar System Exploration. https://solarsystem.nasa.gov/moons/saturn-moons/titan/overview/

- *Voyager - Mission Status*. (2021). NASA JPL. https://voyager.jpl.nasa.gov/mission/status/

- Westre, T. (2019, May 29). *Which World In Our Solar System Has The Most Water?* UPR Utah Public Radio. https://www.upr.org/post/which-world-our-solar-system-has-most-water

- *What Is Gravity? | NASA Space Place – NASA Science for Kids*. (2020, December 17). NASA Space Place. https://spaceplace.nasa.gov/what-is-gravity/en/

- *Which is the largest crater in the Solar System?* (2020, August 5). Space Centre. https://www.spacecentre.nz/resources/faq/solar-system/largest-crater.html

- Whittaker, I. (2021, July 11). *From iron rain on exoplanets to lightning on Jupiter: four examples of alien weather.* Space.Com. https://www.space.com/alien-weather-solar-system-and-exoplanets

- Wikipedia contributors. (2021a, July 25). *Tidal locking.* Wikipedia. https://en.wikipedia.org/wiki/Tidal_locking

- Wikipedia contributors. (2021b, July 26). *South Pole–Aitken basin.* Wikipedia. https://en.wikipedia.org/wiki/South_Pole%E2%80%93Aitken_basin

- (2015, December 25). *Solar System Orbits.* Universe Today. https://www.universetoday.com/37512/solar-system-orbits/

- (2016, July 26). *How Far is Neptune's from the Sun?* Universe Today. https://www.universetoday.com/44572/neptunes-distance-from-the-sun/

Chapter Two: Solar System

- *Ceres.* (2021, February 15). NASA Solar System Exploration. https://solarsystem.nasa.gov/planets/dwarf-planets/ceres/overview/

- E. (2020, December 9). *Which spiral arm of the Milky Way contains our sun? | Space | EarthSky*. EarthSky | Updates on Your Cosmos and World. https://earthsky.org/space/does-our-sun-reside-in-a-spiral-arm-of-the-milky-way-galaxy/

- Gohd, C. (2020, July 19). *Comet NEOWISE: 10 big questions (and answers) about the icy wanderer*. Space.Com. https://www.space.com/comet-neowise-strange-facts.html

- *Haumea*. (2019, December 19). NASA Solar System Exploration. https://solarsystem.nasa.gov/planets/dwarf-planets/haumea/in-depth/

- Hillock, J. (2021, March 17). *20 Interesting Facts About Comets*. The Fact Site. https://www.thefactsite.com/20-interesting-facts-comets/

- *In Depth | 4 Vesta –*. (2019, December 19). NASA Solar System Exploration. https://solarsystem.nasa.gov/asteroids-comets-and-meteors/asteroids/4-vesta/in-depth/

- *In Depth | 433 Eros –*. (2021, June 29). NASA Solar System Exploration. https://solarsystem.nasa.gov/asteroids-comets-and-meteors/asteroids/433-eros/in-depth/

- *JPL Small Body Database Browser*. (2021). NASA JPL. https://ssd.jpl.nasa.gov/sbdb.cgi?sstr=1999J6;cad=1;orb=0;cov=0;log=0#cad

- *The Kuiper Cliff Mystery - Why does the Kuiper Belt Suddenly End?* (2017, June 12). Bibliotecapleyades.Net. https://www.bibliotecapleyades.net/universo/cosmos315.htm

- Lennon, A. (2021, May 15). *Scientists Find Liquid Water Inside Meteorite.* Labroots. https://www.labroots.com/trending/space/20441/scientists-liquid-water-inside-meteorite

- *The Longest Comet Tail Is Breaking Records.* (2021, June 14). MSN News. https://www.msn.com/en-us/news/watch/the-longest-comet-tail-is-breaking-records/vp-AAL1SAg

- *Lost comet - WikiMili, The Free Encyclopedia.* (2021, July 15). WikiMili.Com. https://wikimili.com/en/Lost_comet

- NASA. (2021). *OSIRIS-REx Overview.* https://www.nasa.gov/content/osiris-rex-overview/

- Pappas, S. (2021, July 12). *These are the best astronomy images of 2021.* Space.Com. https://www.space.com/best-astronomy-photographs-2021

- *Pluto's Unusual Orbit | Exploring the Planets | National Air and Space Museum.* (2021). Smithsonian National Air and Space. https://airandspace.si.edu/exhibitions/exploring-the-planets/online/solar-system/pluto/orbit.cfm

- Sounds, S. (2020, October 9). *A huge solar flare burst from the Sun and just missed Earth. . . Now look at the video and feel how lucky we are.* Strange Sounds. https://strangesounds.org/2020/10/a-huge-solar-flare-burst-from-the-sun-and-just-missed-earth-but-look-at-the-video-and-feel-how-lucky-we-are.html

- Wikipedia contributors. (2021, July 21). *List of comets by type.* Wikipedia. https://en.wikipedia.org/wiki/List_of_comets_by_type

Chapter Three: Beyond Our Solar System

- A. (2016, April 2). *Double Cluster (NGC 869 and NGC 884) | Constellation Guide.* Constellation-Guide.Com. https://www.constellation-guide.com/double-cluster/

- A. (2020, November 13). *Messier 57: Ring Nebula.* Messier Objects. https://www.messier-objects.com/messier-57-ring-nebula/

- A., A., A., A., A., A., & A. (2021, June 7). *Emission Nebula | Constellation Guide.* Constellation-Guide.Com. https://www.constellation-guide.com/category/emission-nebula/

- *APOD: 2002 April 16 - Millions of Stars in Omega Centauri.* (2002, April 16). NASA. https://apod.nasa.gov/apod/ap020416.html

- Astronomy Hq, L. (2018). *The Least Massive Stars*. Learn Astronomy HQ. https://learnastronomyhq.com/articles/the-least-massive-stars.html

- *Cosmic Bow Shocks | Science Mission Directorate*. (2018, March 6). NASA. https://science.nasa.gov/science-news/news-articles/cosmic-bow-shocks

- *Dwarf Galaxy | COSMOS*. (2021). Swinburne University of Technology. https://astronomy.swin.edu.au/cosmos/D/dwarf+galaxy

- Frater, J. (2020, June 12). *Top 10 Most Unusual Structures In The Universe*. Listverse. https://listverse.com/2020/06/12/top-10-most-unusual-structures-in-the-universe/

- Frater, J. (2021, May 31). *Ten Astonishing New Discoveries About The Cosmos*. Listverse. https://listverse.com/2021/06/01/ten-astonishing-new-discoveries-about-the-cosmos/

- Gannon, M. (2013, June 10). *Lightweight Galaxy Is the Smallest Ever Found*. Space.Com. https://www.space.com/21500-smallest-dwarf-galaxy-found.html

- Garner, R. (2020, July 21). *Messier 87*. NASA. https://www.nasa.gov/feature/goddard/2017/messier-87/

- *Globular Clusters.* (2021). Hyper Physics Edu. http://hyperphysics.phy-astr.gsu.edu/hbase/Astro/globular.html

- *Hubble Spots a Secluded Starburst Galaxy.* (2016, July 15). NASA. https://www.nasa.gov/image-feature/goddard/2016/hubble-spots-a-secluded-starburst-galaxy/

- *Lifting the Veil on Star Formation in the Orion Nebula | SOFIA Science Center.* (2019, January 7). Sophia Science Center. https://www.sofia.usra.edu/multimedia/science-results-archive/lifting-veil-star-formation-orion-nebula

- *M13: The Great Globular Cluster in Hercules | Science Mission Directorate.* (2017, May 12). NASA. https://science.nasa.gov/m13-great-globular-cluster-hercules

- Ng, A. (2017, December 7). *Milky Way's Most Massive Star Discovered (PHOTOS).* HuffPost. https://www.huffpost.com/entry/milky-ways-most-massive-s_n_449359

- Siegel, E. (2021, May 19). *Why Aren't Astronomers Paying More Attention To UFOs?* Forbes. https://www.forbes.com/sites/startswithabang/2021/05/19/why-arent-astronomers-paying-more-attention-to-ufos/?sh=23af5ee76d8e

- *Stars that are Giants in their Own Right.* (2020, January 10). ThoughtCo. https://www.thoughtco.com/the-largest-star-in-the-universe-3073629

- *The Top 10 Most Massive Stars.* (2020, January 10). ThoughtCo. https://www.thoughtco.com/the-top-most-massive-stars-3073630

- *Types of Binary Stars.* (2021). Australia Telescope National Facility. https://www.atnf.csiro.au/outreach/education/senior/astrophysics/binary_types.html

- Wikipedia contributors. (2021a, July 10). *Large Magellanic Cloud.* Wikipedia. https://en.wikipedia.org/wiki/Large_Magellanic_Cloud

- Wikipedia contributors. (2021b, July 20). *Gravitational lens.* Wikipedia. https://en.wikipedia.org/wiki/Gravitational_lens

- Wikipedia contributors. (2021c, July 21). *Quasar.* Wikipedia. https://en.wikipedia.org/wiki/Quasar

- Wikipedia contributors. (2021d, July 22). *Satellite galaxy.* Wikipedia. https://en.wikipedia.org/wiki/Satellite_galaxy

- Woo, M. (2016, October 20). *The first planet ever discovered around another star.* BBC. http://www.bbc.com/earth/story/20161019-the-first-planet-around-another-star

- (2017, January 4). *What is the Closest Galaxy to the Milky Way?* Universe Today. https://www.universetoday.com/21914/the-closest-galaxy-to-the-milky-way/

Chapter Four: Astronomy

- *10 Things You Might Not Know About Voyager's Famous 'Pale Blue Dot' Photo.* (2020, February 13). NASA Solar System Exploration. https://solarsystem.nasa.gov/news/1175/10-things-you-might-not-know-about-voyagers-famous-pale-blue-dot-photo/

- A. (2014, June 5). *Orion's Belt: Stars, Facts, Location, Myths | Constellation Guide.* Constellation-Guide.Com. https://www.constellation-guide.com/orions-belt/

- A. (2021, May 19). *How many satellites are orbiting the Earth in 2021?* Pixalytics Ltd. https://www.pixalytics.com/satellites-orbiting-2021/

- *altazimuth coordinate system | Encyclopedia.com.* (2021). Encyclopedia.Com. https://www.encyclopedia.com/reference/encyclopedias-almanacs-transcripts-and-maps/altazimuth-coordinate-system

- *asterism.* (2021, July 26). The Merriam-Webster.Com Dictionary. https://www.merriam-webster.com/dictionary/asterism

- *Astronaut/Cosmonaut Statistics*. (2017, March 30). WorldSpaceFlight.Com. https://www.worldspaceflight.com/bios/stats.php

- Baird, D. (2021, July 8). *NASA Laser Communications Innovations: A Timeline*. NASA. https://www.nasa.gov/feature/goddard/2021/nasa-laser-communications-innovations-a-timeline/

- Barnett, B. A. (2019, March 28). *10 Things: Going Interstellar*. NASA Solar System Exploration. https://solarsystem.nasa.gov/news/881/10-things-going-interstellar/

- Garcia, M. (2021, July 12). *International Space Station Facts and Figures*. NASA. https://www.nasa.gov/feature/facts-and-figures/

- Gohd, C. (2021, July 12). *Dozens of starless "rogue" alien planets possibly spotted*. Space.Com. https://www.space.com/possible-rogue-alien-planets-detected-kepler

- Hall, S. (2021, July 14). *China has landed its first rover on Mars — here's what happens next*. BBC-Edition. https://bbc-edition.com/space/2021/07/14/china-has-landed-its-first-rover-on-mars-heres-what-happens-next/

- *How Does Ancient Mayan Astronomy Portray the Sun, Moon and Planets?* (2019, July 24). ThoughtCo. https://www.thoughtco.com/ancient-maya-astronomy-2136314

- *How many astronauts have died in space?* (2019, October 7). Astronomy.Com. https://astronomy.com/news/2019/10/how-many-astronauts-have-died-in-space

- *In Depth | Deep Impact (EPOXI)* –. (2019, July 24). NASA Solar System Exploration. https://solarsystem.nasa.gov/missions/deep-impact-epoxi/in-depth/

- *In Depth | James Webb Space Telescope* –. (2019, January 11). NASA Solar System Exploration. https://solarsystem.nasa.gov/missions/james-webb-space-telescope/in-depth/

- *In Depth | Voyager 2* –. (2021, February 4). NASA Solar System Exploration. https://solarsystem.nasa.gov/missions/voyager-2/in-depth/

- International Dark-Sky Association. (2021, July 13). *International Dark Sky Parks*. https://www.darksky.org/our-work/conservation/idsp/parks/

- Johnson, M. (2020, October 28). *20 Breakthroughs from 20 Years of Science aboard the ISS*. NASA. https://www.nasa.gov/mission_pages/station/research/news/iss-20-years-20-breakthroughs/

- Malik, T. (2009, August 3). *Tool Bag Lost In Space Meets Fiery End*. Space.Com. https://www.space.com/7088-tool-bag-lost-space-meets-fiery.html

- NASA. (2021, February 26). *NASA at Home -- Virtual Tours and Apps.* https://www.nasa.gov/nasa-at-home-virtual-tours-and-augmented-reality/

- *NASA - The First Person on the Moon.* (2008, January 16). NASA. https://www.nasa.gov/audience/forstudents/k-4/stories/first-person-on-moon.html

- News18. (2020, October 7). *The Last Time All Humans Were on Earth Together Was Almost Two Decades Ago.* https://www.news18.com/news/tech/the-last-time-all-humans-were-on-earth-together-was-almost-two-decades-ago-2940557.html

- PetaPixel. (2013, February 11). *The Family Photo That Was Left on the Surface of the Moon.* https://petapixel.com/2013/02/11/the-family-photo-that-made-it-to-the-moon/

- Roos, D. (2021, July 19). *Apollo 11 Moon Landing Timeline: From Liftoff to Splashdown.* HISTORY. https://www.history.com/news/apollo-11-moon-landing-timeline

- *Voyager - Mission Status.* (2021). NASA JPL. https://voyager.jpl.nasa.gov/mission/status/

- Wall, M. (2016, July 6). *China Finishes Building World's Largest Radio Telescope.* Space.Com. https://www.space.com/33357-china-largest-radio-telescope-alien-life.html

- *What should I be looking for when buying my first telescope? | Space Facts – Astronomy, the Solar System & Outer Space | All About Space Magazine.* (2013, December 6). Space Answers. https://www.spaceanswers.com/astronomy/what-should-i-be-looking-for-when-buying-my-first-telescope/

- Wikipedia contributors. (2021a, July 13). *Heliocentrism.* Wikipedia. https://en.wikipedia.org/wiki/Heliocentrism

- Wikipedia contributors. (2021b, July 13). *William Herschel.* Wikipedia. https://en.wikipedia.org/wiki/William_Herschel

- Wikipedia contributors. (2021c, July 24). *Service structure.* Wikipedia. https://en.wikipedia.org/wiki/Service_structure#White_room

- Wikipedia contributors. (2021d, July 25). *Isaac Newton.* Wikipedia. https://en.wikipedia.org/wiki/Isaac_Newton

- World History Edu. (2021, May 16). *12 Important Facts about Ancient Mesopotamia.* https://www.worldhistoryedu.com/12-interesting-facts-about-ancient-mesopotamia/

- (2019, March 19). *35th Anniversary of the Voyager 1 Saturn Flyby.* NASA Solar System Exploration. https://solarsystem.nasa.gov/news/12963/35th-anniversary-of-the-voyager-1-saturn-flyby/

www.ingramcontent.com/pod-product-compliance
Lightning Source LLC
Chambersburg PA
LSHW071955260326
914CB00004B/805